Visual Saliency: From Pixel-Level to Object-Level Analysis

Jianming Zhang • Filip Malmberg • Stan Sclaroff

Visual Saliency: From Pixel-Level to Object-Level Analysis

Jianming Zhang
Adobe Inc.
San Jose, CA, USA

Filip Malmberg
Centre for Image Analysis
Uppsala University
Uppsala, Uppsala Län, Sweden

Stan Sclaroff
Department of Computer Science
Boston University
Boston, MA, USA

ISBN 978-3-030-04830-3   ISBN 978-3-030-04831-0   (eBook)
https://doi.org/10.1007/978-3-030-04831-0

Library of Congress Control Number: 2018965600

This Springer imprint is published by the registered company Springer Nature Switzerland AG.
The registered company address is: Gewerbestrasse 11, 6330 Cham, Switzerland

Contents

Chapter 1
Overview

Visual saliency computation is about detecting and understanding pertinent regions and elements in a visual scene. Given limited computational resources, the human visual system relies on saliency computation to quickly grasp important information from the excessive input from the visual world [189]. Modeling visual saliency computation can help computer vision systems to filter out irrelevant information and thus make them fast and smart. For example, saliency detection methods have been proposed to predict where people look [87], delineate between foreground regions and the background [1], and localize dominant objects in images [112]. These techniques have been used in many computer vision applications, e.g. image segmentation [69], object recognition [150], visual tracking [118], gaze estimation [169], action recognition [125], and so on.

In this book, we present methods for both traditional and emerging saliency computation tasks, ranging from classical low-level tasks such as pixel-level saliency detection to emerging object-level tasks such as subitizing and salient object detection. For low-level tasks, we focus on pixel-level image processing approaches based on efficient distance transform. For object-level tasks, we propose data-driven methods using deep convolutional neural networks. The book includes both empirical and theoretic studies, together with implementation details of the proposed methods. The rest of this chapter will introduce the background of those saliency computation tasks and the outline of this book.

1.1 Pixel-Level Saliency Detection

Early works in visual saliency computational modelling [79, 95] aim at computing a saliency/attention map that topographically represents humans' attentional priority when they view a scene. Computing such saliency maps is formulated as assigning importance levels to each pixel of a digital image. At the beginning of this line

J. Zhang et al., *Visual Saliency: From Pixel-Level to Object-Level Analysis*,
https://doi.org/10.1007/978-3-030-04831-0_1

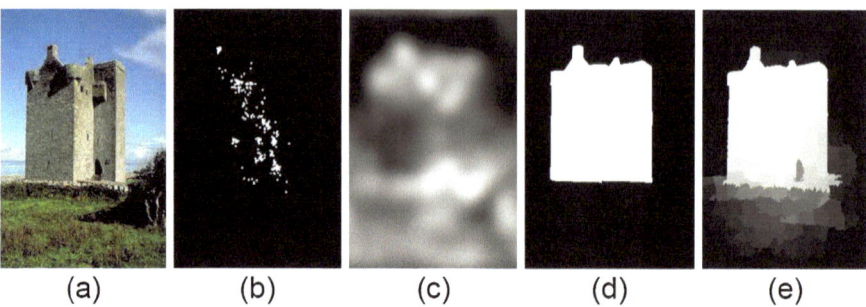

Fig. 1.1 (**a**) A sample image from the DUT-Omron dataset [194]. (**b**) The ground truth for eye fixation prediction. Each white spot on the map is an eye fixation position of some participant in the free-viewing experiment. (**c**) The saliency map by a state-of-the-art method for eye fixation prediction [61]. (**d**) The ground truth for salient region detection. (**e**) The saliency map by a state-of-the-art method for salient object detection [84]

of research, the saliency values are represented based on the likelihood of human eye gaze. Therefore, this type of saliency models are often named as eye fixation prediction models. Later, motivated by image segmentation, people started to compute saliency maps that delineate dominant objects from the background. This type of saliency detection task is named interchangeably as salient object detection, salient object segmentation, or salient region detection. In this book, we will refer to it as salient region detection, since most of the existing works of this topic are still based on very low-level image processing algorithms without semantic object-level understanding.

Figure 1.1 shows the difference between the two pixel-level saliency tasks mentioned above. Figure 1.1b, c is the ground truth eye fixation map and the predicted saliency map, respectively. The eye fixation data is collected by eye tracking devices from multiple subjects when they free-view the image, and the data is expected to have a significant amount of inter-subject variance. By aggregating the eye fixations of different subjects, the ground truth map shows that the human eye gaze is often focused on salient regions or objects, but there is still a lot of uncertainty about the exact positions of those eye fixations. As a result, the predicted saliency map tends to be very blurry. In contrast, the ground truth map for salient region detection is simply a binary foreground mask for the dominant object in the image, and thus has much less uncertainty (see Fig. 1.1c) when there is a well-defined dominant object. The predicted map needs to uniformly highlight the regions of dominant objects with precise object boundary details, as shown in Fig. 1.1d.

These two pixel-level saliency tasks are useful in many applications. Predicting human eye fixation is useful in applications related to human–computer interaction and graphics, e.g. gaze estimation [169], eye tracker calibration [168], non-photorealistic rendering [47], stereoscopic disparity manipulations [92], image retargeting [146], photo quality assessment [131], etc. Other computer vision tasks such as action recognition, tracking, and object detection can also benefit from this task by analyzing relevant regions indicated by human eye fixation [118, 125, 150]. Salient region detection methods can automatically generate saliency maps to extract dominant objects, and thus this task is useful for automatic image segmentation [32] and a lot of photo editing applications that need such segmentation [30, 37, 76].

1.2   Object-Level Saliency Computation

Object-level saliency computation is a relatively new topic. By object level, we mean that the analysis should be performed based on the understanding of object instances. One of the basic problems, for example, is to predict the existence of salient object(s) in an image. Existence prediction leads to the differentiation between object-centric images and scene-centric images, which is a very general and fundamental attribute for image understanding.

In this book, we propose a new problem of visual saliency computation, called salient object subitizing (SOS), which is to predict not only the existence but also the number of salient objects in an image using holistic cues, without the need to localize them. This task is inspired by humans' subitizing ability to quickly and accurately tell the number of items within the subitizing range (1–4) [90]. Because the appearance and size of salient objects can vary dramatically from category to category, and from image to image, the SOS problem poses very different challenges than traditional object counting problems [3, 129].

Knowing the existence and the number of salient objects without the expensive detection process can enable a machine vision system to select different processing pipelines at an early stage, making the vision system more intelligent and reducing computational cost. Furthermore, differentiating between scenes with zero, a single and multiple salient objects can also facilitate applications such as image retrieval, iconic image detection [11], image thumbnailing [38], robot vision [152], egocentric video summarization [104], snap point prediction [191], etc.

Besides the subitizing task, detecting generic salient object instances in unconstrained images, which may contain multiple salient objects or no salient object, is also a fundamental problem (see examples in Fig. 1.2). We refer to this task as unconstrained salient object detection. Solving this problem entails generating a compact set of detection windows that matches the number and the locations of salient objects. Detecting each salient object (or reporting that no salient object is present) can be very helpful in the weakly supervised or unsupervised learning scenario [31, 89, 207], where object appearance models are to be learned with no instance level annotation.

The unconstrained salient object detection task arguably solves the problem of salient object subitizing, as the existence and the number of salient objects can be derived from the detection result. However, salient object subitizing can be solved by much faster and lighter-weight models, since it does not require accurate

Fig. 1.2 Examples of unconstrained salient object detection. Note that for the input image in the right column, there is no dominant object

localization of salient objects. Moreover, for training a subitizing model, the labor cost for annotation is also much less than the detection model. Therefore, one can expect that the subitizing model can do a better job in detecting the existence and counting the number for salient objects given the same amount of labor and computation resources. This will also be empirically verified in this book.

1.3 Book Outline

The book will cover each of the saliency computation tasks mentioned above and present computational approaches as well as empirical and theoretic studies. Here we provide a summary of the main content.

1.3.1 Boolean Map Saliency for Eye Fixation Prediction

A majority of existing eye fixation prediction models are based on the contrast and the rarity properties of local image patches, e.g. [13, 19, 79]. However, these local image properties have limited ability to model some global perceptual phenomena [96] known to be relevant to the deployment of visual attention. One such global perception mechanism is figure-ground segregation. Several factors are likely to influence figure-ground segregation, e.g. size, surroundedness, convexity, and symmetry [133]. As Gestalt psychological studies suggest, figures are more likely to be attended to than background elements [126, 145] and the figure-ground assignment can occur without focal attention [94]. Neuroscience findings also show that certain responses in monkey and human brains involved in shape perception are critically dependent on figure-ground assignment [10, 98], indicating that this process may start early in the visual system.

In the first part of this book, we explore the usefulness of the surroundedness cue for eye fixation detection [201, 202]. We propose a simple, training-free, and computationally efficient Boolean map saliency model (BMS). Our model uses basic image processing operations to find surrounded regions in binary maps which are generated by randomly thresholding the color channels of an image. The saliency map is computed based on the probability that a pixel belongs to a surrounded region in a sampled binary map. Despite its simplicity, BMS consistently achieves state-of-the-art performance across all the testing datasets. Regarding the efficiency, BMS can be configured to run at about 100 FPS with only a little drop in performance, which makes it quite suitable for many time-critical applications.

1.3.2   A Distance Transform Perspective

We provide an explanation of the BMS algorithm in a perspective of image distance transform [120, 202]. First, we propose a novel distance function, the Boolean map distance (BMD), that defines the distance between two elements in an image based on the probability that they belong to different components after thresholding the image by a randomly selected threshold value. We show that the BMS algorithm is a straightforward implementation to compute the Boolean map distance of each pixel to the image border.

Then we draw a connection between the Boolean map distance and the minimum barrier distance (MBD) [166]. We prove that the Boolean map distance gives a lower bound approximation of the minimum barrier distance. As such it shares many of the favorable properties of the MBD discovered in [40, 166], while offering some additional advantages such as more efficient distance transform computation and straightforward extension to multi-channel images. These analyses provide insight into why and how BMS can capture the surroundedness cue via Boolean maps.

Finally, we discuss efficient algorithms for computing the Boolean map distance and the minimum barrier distance. In the next chapter, we propose a fast raster-scanning algorithm to approximate BMD and MBD, and show how to use that for real-time salient region detection.

1.3.3   Efficient Distance Transform for Salient Region Detection

Due to the emerging applications on mobile devices and large-scale datasets, a desirable salient region detection method should not only output high quality saliency maps, but should also be highly computationally efficient. In this chapter, we address both the quality and speed challenges for salient region detection using an efficient distance transform algorithm.

The surroundedness prior, also known as the *image boundary connectivity prior*, assumes that background regions are usually connected to the image borders. This prior is shown to be effective for salient region detection [187, 194, 201, 208]. To leverage this prior, previous methods, geodesic-distance-based [187, 208] or diffusion-based [84, 194], rely on a region abstraction step to extract superpixels. The superpixel representation helps remove irrelevant images details, and/or makes these models computationally feasible. However, this region abstraction step also becomes a speed bottleneck for this type of methods.

To boost the speed, we propose a method to exploit the image boundary connectivity prior without region abstraction [203]. Inspired by the connection between our BMS eye fixation prediction method and the *minimum barrier distance* (MBD) [40, 166], we use the MBD [40, 166] to measure a pixel's connectivity to the image boundary. Compared with the widely used geodesic distance, the MBD is much more robust to pixel value fluctuation. Since the exact algorithm for the MBD transform is not very efficient, we present FastMBD, a fast raster-scanning algorithm for the MBD transform, which provides a good approximation of the MBD transform in milliseconds, being two orders of magnitude faster than the exact algorithm [40]. The proposed salient region detection method runs at about 80 FPS using a single thread, and achieves comparable or better performance than the leading methods on four benchmark datasets. Compared with methods with similar speed, our method gives significantly better performance.

1.3.4 Salient Object Subitizing

We introduce a new computer vision task, salient object subitizing (SOS), to estimate the existence and the number of salient objects in a scene [199, 200]. To study this problem, we present a salient object subitizing image dataset of about 14K everyday images. The number of salient objects in each image was annotated by Amazon Mechanical Turk (AMT) workers. The resulting annotations from the AMT workers were analyzed in a more controlled offline setting; this analysis showed a high inter-subject consistency in subitizing salient objects in the collected images.

We formulate the SOS problem as an image classification task, and aim to develop a method to quickly and accurately predict the existence and the number of generic salient objects in everyday images. We propose to use an end-to-end trained deep convolutional neural network (CNN) model for our task, and show that an implementation of our method achieves very promising performance. In particular, the CNN-based subitizing model can approach human performance in identifying images with no salient object and with a single salient object. To further improve the training of the CNN SOS model, we propose a method to leverage synthetic images. Moreover, we demonstrate the application of our CNN-based SOS method in salient object detection and image retrieval.

1.3.5   Unconstrained Salient Object Detection

We introduce another new task, called unconstrained salient object detection. Many previous so-called "salient object detection" methods [1, 12, 34, 85, 113, 158] only solve the task of salient region detection, i.e. generating a dense foreground mask (saliency map). These methods do not individuate each object and assume the existence of salient objects. In contrast, we present a salient object detection system that directly outputs a compact set of instance detection windows for an unconstrained image, which may or may not contain salient objects. Our system leverages the high expressiveness of a convolutional neural network (CNN) model to generate a set of scored salient object proposals for an image. Inspired by the attention-based mechanisms of [8, 102, 127], we propose an adaptive region sampling method to make our CNN model "look closer" at promising images regions, which substantially increases the detection rate. The obtained proposals are then filtered to produce a compact detection set.

A key difference between salient object detection and object class detection is that saliency greatly depends on the surrounding context. Therefore, the salient object proposal scores estimated on local image regions can be inconsistent with the ones estimated on the global scale. This intrinsic property of saliency detection makes our proposal filtering process very challenging. Using the common greedy non-maximum suppression (NMS) method often leads to suboptimal results for our proposals. To attack this problem, we propose a subset optimization formulation based on the *maximum a posteriori* (MAP) principle, which jointly optimizes the number and the locations of detection windows. The effectiveness of our optimization formulation is validated on three benchmark datasets, where our formulation attains about 15% relative improvement in average precision (AP) over the NMS approach. Moreover, our method also attains about 15–35% relative improvement in AP over previous methods on these datasets.

Part I
Pixel-Level Saliency

Chapter 2
Boolean Map Saliency: A Surprisingly Simple Method

In this chapter, we propose a simple yet powerful saliency detection model for eye fixation prediction based on the surroundedness cue for figure-ground segregation. The essence of surroundedness is the enclosure topological relationship between different visual components. This kind of topological relationship is invariant under homeomorphisms; thus, it is a quite fundamental property of a scene, regardless of the scale or the shape of the visual content. It is also worth noting that the topological status of a scene has long been identified as one of the probable attributes that guide the deployment of visual attention [189].

To demonstrate the strength of the surroundedness cue for saliency detection, we propose a simple, training-free, and computationally efficient Boolean map based saliency (BMS) model. In our formulation, an image is characterized by a set of randomly sampled Boolean maps. For each Boolean map, an attention map is efficiently computed by binary image processing techniques to activate regions with closed outer contours. Then attention maps are averaged into a mean attention map, which is further post-processed to suit the purpose of eye fixation prediction.

Figure 2.1 shows an example that the surroundedness cue for figure-ground segregation can help in saliency detection. A test image along with eye tracking data is displayed in Fig. 2.1a. The bird in the image is naturally perceived as the foreground and the rest as the background, which is in agreement with the enclosure relationship between the bird and the sky. The eye fixations are concentrated on the bird, corresponding well to this figure-ground assignment. However, without the awareness of this global structure, rarity based models [13, 19] falsely assign high saliency values to the boundary area between the trees and the sky, due to the rarity of high contrast regions in natural images. In contrast, by leveraging the surroundedness cue for figure-ground segregation, our model is less responsive to the edges and cluttered areas in the background (Fig. 2.1b).

BMS is extensively evaluated on seven eye tracking datasets, comparing with ten state-of-the-art saliency models under two evaluation metrics. Detailed speed performance and component analyses are also provided. In our experiments, most

J. Zhang et al., *Visual Saliency: From Pixel-Level to Object-Level Analysis*, https://doi.org/10.1007/978-3-030-04831-0_2

Fig. 2.1 (**a**) Image from the
MIT dataset [87] (left) and its
eye tracking data (right). (**b**)
Saliency maps estimated by
(from left to right) AIM [19],
LG [13], and our method.
AIM and LG measure an
image patch's saliency based
on its rarity. Our method,
based on global structural
information, is less
responsive to the elements in
the background

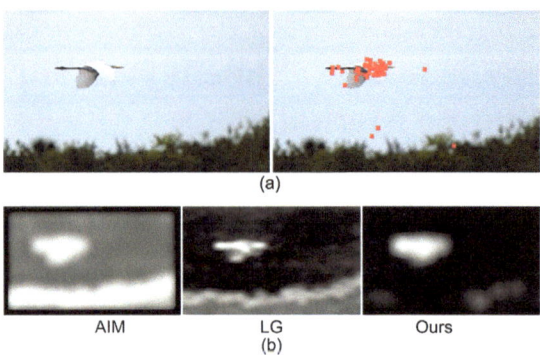

of the competing models critically rely on off-line training or multi-scale processing, while BMS uses neither of them. Despite its simplicity, BMS consistently
achieves state-of-the-art performance across all the testing datasets. Regarding
the efficiency, BMS can be configured to run at about 100 FPS with only a
little drop in performance, which makes it quite suitable for many time-critical
applications.

2.1 Related Work

There have been many eye fixation models based on a variety of principles and
assumptions. We review some typical approaches to this problem.

Rarity/Contrast Based Models A majority of the previous eye fixation models use
center-surround contrast or image statistics to identify salient patches that are complex (local complexity/contrast) or rare in their appearance (rarity/improbability).
Center-surround contrast is used by [54, 79, 156, 175] for eye fixation prediction.
The contrast and rarity of an image region are also widely exploited for eye fixation
prediction via information theoretic models [19, 205], Bayesian probabilistic models
[78], graphical models [71], color co-occurrence histogram [114], and feature vector
differences [27, 53, 61].

Spectral Analysis Models Another family of eye fixation prediction models is
based on spectral domain analysis [74, 75, 107, 153]. In [107] it is argued that
some previous spectral analysis based methods are equivalent to a local gradient
operator plus Gaussian blurring, and thus cannot detect large salient regions very
well. To overcome this limitation, a method based on spectral scale-space analysis
is proposed by [107].

Learning Based Models Some models employ machine learning to learn eye
fixation patterns. Kienzel et al. [93] learn a kernel support vector machine (SVM)
based on eye tracking data. Judd et al. [87] train an SVM using a combination of low,

middle, and high level features. In [101, 183], convolutional neural network models are leveraged for eye fixation prediction. A linear weighting function is learned by [70] to combine different types of eye fixation models.

Unlike the previous approaches, our proposed BMS formulation does not rely on center-surround filtering, statistical analysis of features, spectral transforms, off-line learning, or multi-scale processing. Instead, it makes use of simple image processing operations to leverage the topological structural cue, which is scale-invariant and known to have a strong influence on visual attention [29, 189].

2.2   Boolean Map Based Saliency

We start with a general description of our basic formulation. We borrow the *Boolean map* concept that was put forward in the Boolean map theory of visual attention [77], where an observer's momentary conscious awareness of a scene can be represented by a Boolean map. We assume that Boolean maps in BMS is generated by sampling from a distribution function $F(\mathcal{B}|\mathcal{I})$ conditioned on the input image $\mathcal{I}$, and the influence of a Boolean map $\mathcal{B}$ on visual attention can be represented by an *attention map* $\mathcal{A}(\mathcal{B})$, which highlights regions on $\mathcal{B}$ that attract visual attention. Then the saliency is modeled by the mean attention map $\bar{\mathcal{A}}$ over randomly generated Boolean maps:

$$\bar{\mathcal{A}} = \int \mathcal{A}(\mathcal{B}) dF(\mathcal{B}|\mathcal{I}), \tag{2.1}$$

where $\bar{\mathcal{A}}$ can be further post-processed to suit the purpose of eye fixation prediction.

In our formulation, computing the attention map $A(\mathcal{B})$ for a Boolean map requires two steps: an *activation* step and a *normalization* step. In the activation step, a Boolean *activation map* $\mathcal{M}(\mathcal{B})$ is produced by removing unsurrounded regions on the Boolean map; in the normalization step, the attention map is computed by normalizing the activation map to emphasize those rare activated regions. The pipeline of the computation of the mean attention map is illustrated in Fig. 2.2.

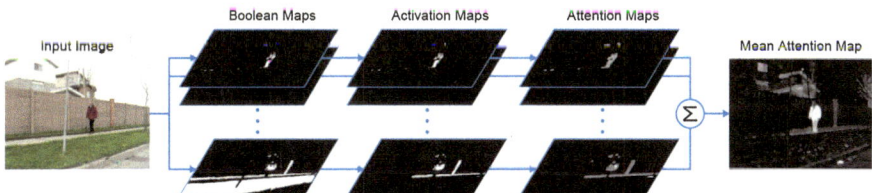

Fig. 2.2 The pipeline of BMS. An image is first represented by a set of randomly generated Boolean maps. For each Boolean map, a binary activation map is produced by suppressing unsurrounded regions. Then a real-valued attention map is obtained by normalizing the activation map. At last, attention maps are linearly combined

2.2.1 Boolean Map Generation

BMS samples a set of Boolean maps by randomly thresholding the input image's feature maps, according to the prior distributions over the feature channels and the threshold:

$$\mathcal{B}_i = \textbf{THRESH}(\phi(\mathcal{I}), \theta), \tag{2.2}$$

$$\phi \sim F_\phi, \theta \sim F_\theta^\phi.$$

The function $\textbf{THRESH}(., \theta)$ assigns 1 to a pixel if its value on the input map is greater than θ, and 0 otherwise. $\phi(\mathcal{I})$ denotes a feature map of $\mathcal{I}$, whose values, without loss of generality, are assumed to range between 0 and 1. F_ϕ denotes the prior distribution function for feature channel sampling, and F_θ^ϕ denotes the prior distribution function for the threshold sampling on the feature channel ϕ. Feature channels can consist of multiple features such as color, intensity, depth, motion, etc. In this work, we demonstrate the proposed formulation in an implementation using only color channels of images. Note that although feature maps generated by various image filters are widely used in previous rarity/contrast based saliency models [8, 9, 30], our preliminary study shows that this feature maps of this type are not suitable for measuring surroundedness in our formulation, because they tend to lose the topological structure of the scene by only sparsely highlighting certain local patterns (e.g., edges and corners).

According to [52, Theorem 2.1], we have the following fact.

Proposition 2.1 *Let F_θ be a continuous cumulative distribution function of variable θ, and U be a random variable with uniform distribution over $[0, 1]$. Then*

$$P(\theta \le x) = P(U \le F_\theta(x)), \ \forall x \in \mathbb{R}.$$

It means that we can get equivalent sampling of Boolean maps from a feature map $\phi(\mathcal{I})$ by first re-mapping the values of $\phi(\mathcal{I})$ using F_θ^ϕ, and then sampling a threshold from a uniform distribution over $[0, 1]$. Thus, without loss of generality, we can always assume that the threshold θ is drawn from a uniform distribution over $[0, 1]$. As a result, the distribution of generated Boolean maps is now only dependent on the specific parametrization of the feature space and the prior distribution for the feature channel selection.

If we want to further simplify the sampling process by assuming equal importance of different color channels, i.e. F_ϕ is uniform, then the color space should have independent channels and the distance metric on different channels should be comparable. Following this intuition, we propose a color whitening step to rectify a color space before sampling the Boolean maps. Let $\mathbf{x}_i = (x_i^1, x_i^2, x_i^3)^T$ be the 3-D color vector of a pixel indexed by i. Given an image, we first compute the color mean and the color covariance matrix as follows:

$$\bar{\mathbf{x}} = \frac{1}{n} \sum_i \mathbf{x}_i, \tag{2.3}$$

$$\mathbf{Q} = \frac{1}{n} \sum_i \mathbf{x}_i \mathbf{x}_i^T - \bar{\mathbf{x}}\bar{\mathbf{x}}^T, \tag{2.4}$$

where n is the number of pixels on the given image. Then we transform the color space by

$$\mathbf{y}_i = (\mathbf{Q} + \lambda \mathbf{I})^{-\frac{1}{2}} \cdot \mathbf{x}_i, \tag{2.5}$$

where $\mathbf{y}_i$ is the color vector in the whitened space. $\mathbf{I}$ is the identity matrix and λ serves as regularization parameter to avoid degeneracy. Note that this whitening process is not limited to color space, and it can help de-correlate and normalize feature channels of multiple features such as color, depth, motion, etc. Feature space whitening has been used for saliency detection in [61, 142], but for a different purpose. In [61, 142], a pixel's saliency is directly measured by its distance to the sample mean in the whitened space.

In summary, to generate Boolean maps for an image, we first do color space whitening according to Eqs. 2.3–2.5. Then we simply enumerate the color channels and sample the threshold θ at a fixed step size δ within the range of that channel. Note that in the limit, a fixed-step sampling is equivalent to the uniform sampling.

2.2.2 Attention Map Computation

Given a Boolean map $\mathcal{B}$, BMS computes the attention map $\mathcal{A}(\mathcal{B})$ by first activating the surrounded regions on $\mathcal{B}$, and then normalizing the resultant activation map $\mathcal{M}(\mathcal{B})$ to further emphasize rare regions. We now describe these two steps in detail.

Activation

On a Boolean map $\mathcal{B}$, the pixels are separated into two complementary sets: the white set $\mathbf{C}^+ := \{i : \mathcal{B}(i) = 1\}$ and the black set $\mathbf{C}^- := \{i : \mathcal{B}(i) = 0\}$, where i denotes the pixel index and $\mathcal{B}(i)$ denotes the Boolean value of pixel i on $\mathcal{B}$. Intuitively, a white (black) pixel is surrounded iff it is enclosed by the black (white) set, i.e. it lies in a hole of the black (white) set. Formally, the surroundedness can be defined based on a pixel's connectivity to the image border pixels.

Definition 2.2 On a Boolean map $\mathcal{B}$, a pixel i is *surrounded* if there exists no path in $\mathbf{C}^+$ or $\mathbf{C}^-$ that joins i and any image border pixel.

Here, a path is a sequence of pixels in which any pair of consecutive pixels are adjacent. On a 2D image, we consider 4-adjacency or 8-adjacency. It follows that a pixel is surrounded iff it is not in a connected component of $\mathbf{C}^+$ or $\mathbf{C}^-$ that contains any image border pixels. Therefore, pixels that are not surrounded can be efficiently masked out by a flood fill algorithm using all image border pixels as the seeds. The resultant activation map $\mathcal{M}(\mathcal{B})$ has 1s for all the surrounded pixels and 0s for the rest.

Moreover, we have the following simple fact.

Proposition 2.3 *Let $\neg\mathcal{B}$ denote the inversion of a Boolean map $\mathcal{B}$. A pixel is surrounded in $\mathcal{B}$ iff it is surrounded in $\neg\mathcal{B}$, i.e. $\mathcal{M}(\mathcal{B})$ equals $\mathcal{M}(\neg\mathcal{B})$.*

Therefore, we do not need to activate the inverted copy of a Boolean map. In our previous version of BMS [201], activation is done also for the inverted copy of a Boolean map, because an opening operation was applied before activation, and thus the activation maps of a Boolean map and its inverted version could be slightly different. The opening operation, which we find quite unimportant, is removed from our improved formulation of BMS.

Normalization

The resultant activation maps need to be normalized, so that activation maps with small concentrated active areas will receive more emphasis. Various normalization schemes have been proposed in previous works [71, 79]. In BMS, we first split an activation map $\mathcal{M}(\mathcal{B})$ into two sub-activation maps:

$$\mathcal{M}^+(\mathcal{B}) = \mathcal{M}(\mathcal{B}) \wedge \mathcal{B}, \tag{2.6}$$

$$\mathcal{M}^-(\mathcal{B}) = \mathcal{M}(\mathcal{B}) \wedge \neg\mathcal{B}, \tag{2.7}$$

Algorithm 1 $\bar{\mathcal{A}} = \textbf{BMS}(\mathcal{I})$

1: $\bar{\mathcal{A}} \leftarrow \textbf{ZEROS}(\mathcal{I}.\texttt{size}())$;
2: do feature space whitening by Eqs. 2.3–2.5;
3: **for all** feature maps $\phi_k(\mathcal{I}) :\ k = 1, 2 \cdots N$ **do**
4:    **for** $\theta = \min_i \phi_k(\mathcal{I})(i) : \delta : \max_i \phi_k(\mathcal{I})(i)$ **do**
5:       $\mathcal{B} \leftarrow \textbf{THRESH}(\phi_k(\mathcal{I}), \theta)$;
6:       compute $\mathcal{M}(\mathcal{B})$ according to Sect. 2.2.2;
7:       compute $\mathcal{A}(\mathcal{B})$ according to Eqs. 2.6–2.9;
8:       $\bar{\mathcal{A}} \leftarrow \bar{\mathcal{A}} + \mathcal{A}(\mathcal{B})$;
9:    **end for**
10: **end for**
11: $\bar{\mathcal{A}} \leftarrow \bar{\mathcal{A}}/\max_i \mathcal{A}(i)$;
12: **return** $\bar{\mathcal{A}}$;

---

where $\wedge(\cdot, \cdot)$ is the pixel-wise Boolean conjunction operation. Note that activation maps are not split in the previous version of BMS [201]. $\mathcal{M}^+(\mathcal{B})$ and $\mathcal{M}^-(\mathcal{B})$ represent the selected and surrounded regions on $\mathcal{B}$ and $\neg\mathcal{B}$, respectively. An intuitive interpretation of these sub-activation maps is that $\mathcal{M}^+(\mathcal{B})$ activates

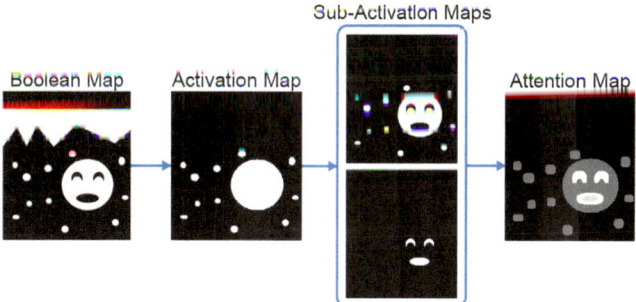

Fig. 2.3 The pipeline of the attention map computation. See the text for details

the surrounded peaks above the corresponding threshold and $\mathcal{M}^-(\mathcal{B})$ activates the surrounded valleys below it. In this sense, normalization can be regarded as a way to emphasize the regions of rare topographic features.

After the above steps, BMS uses simple L2-normalization to emphasize attention maps with small active areas. Compared with L1-normalization, L2-normalization is less sensitive to activation maps with extremely small active areas, which will otherwise dominate the fusion process. To further penalize sub-activation maps with small, scattered active areas, we dilate the sub-activation maps with a kernel of width ω before normalization. Formally, we have

$$\mathcal{A}^\circ(\mathcal{B}) = \mathcal{M}^\circ(\mathcal{B}) \oplus \mathcal{K}_\omega : \circ = +, -; \tag{2.8}$$

$$\mathcal{A}(\mathcal{B}) = \frac{\mathcal{A}^+(\mathcal{B})}{\parallel \mathcal{A}^+(\mathcal{B}) \parallel_2} + \frac{\mathcal{A}^-(\mathcal{B})}{\parallel \mathcal{A}^-(\mathcal{B}) \parallel_2}, \tag{2.9}$$

where $\mathcal{K}_\omega$ is the square dilation kernel of width ω. The pipeline of the attention map computation is shown in Fig. 2.3.

Finally, all the attention maps are averaged into a mean attention map $\bar{\mathcal{A}}$. The complete algorithm of BMS is summarized in Algorithm 1.

2.3   Experiments

Implementation Details   Each input image is first resized to 400 pixels in its largest dimension. We use the CIE Lab color space, and the whitened color channels are transformed to 8-bit images. The threshold sample step δ is set to 8 and the dilation kernel width ω is fixed at 7. We post-process $\bar{A}$ to produce the saliency map S by Gaussian blurring with standard deviation (STD) σ. However, strong Gaussian blur will remove small peaks on the mean attention map, which is sometimes undesirable. To control for this factor, we use a dilation operation with kernel width κ before Gaussian blur. We do not find this dilation operation improves the performance of other compared methods. By experiment, we have found setting σ to 9 and κ to 9 usually works well. We fix these parameters in the following experiments. The source code is available on our website.[1]

2.3.1 Datasets

We use seven benchmark eye tracking datasets: MIT [87], Toronto [19], Kootstra [97], Cerf [22], ImgSal [107], SBU-VOC [196], and DUT-O [194]. These datasets are available on the authors' websites. Some statistics and features of these datasets are summarized in Table 2.1. These datasets differ in many aspects, such as the number of participants, number of test images, type of stimuli, experimental settings, post-processing, etc. The diversity of these datasets ensures a comprehensive evaluation of our model.

In particular, the MIT, Toronto, Kootstra, and ImgSal datasets have a relatively large number of participants for the collection of the eye tracking data, while the MIT, DUT-O, and the SBU-VOC datasets have a large number of test images. Among these datasets, the MIT and Toronto datasets are the most widely used ones for evaluating eye fixation prediction methods. The Cerf dataset has a special focus on images with human faces. The ImgSal and DUT-O datasets also provide segmentation ground truth for salient object detections, and a large portion of their

Table 2.1 Eye tracking datasets

Datasets	#Images	#Viewers	Description
MIT [87]	1003	15	Daily life indoor and outdoor pictures; portraits
Toronto [19]	120	20	A large portion of images do not contain particular regions of interest
Kootstra [97]	100	31	Five categories of images: 12 animals, 12 cars and streets, 16 buildings, 20 flowers and plants, and 40 natural scenes
Cerf [22]	181	8	The objects of interest are mostly faces, together with some small objects like cell phone, toys, etc.
ImgSal [107]	235	21	Six categories: 50/80/60 with large/medium/small salient regions; 15 with clustering background; 15 with repeating distracters; 15 with both large and small salient regions
SBU-VOC [196]	1000	3	Selected images from the Pascal VOC2008 dataset. A mixture of different kinds of daily life pictures, including scenery, animals, portraits, objects, etc.
DUT-O [194]	5168	5	Most of the images contain salient objects. Objects are usually centered in these images

---

[1] http://www.cs.bu.edu/groups/ivc/software/BMS/.

test images contain dominant salient objects. The SUB-VOC dataset is proposed for analyzing the utility of eye fixations for object detection and its test images usually contain people and multiple objects. Moreover, a post-processing step is applied in the DUT-O dataset to remove outlier eye fixations that do not lie on a meaningful object, which leads to a higher center bias of its ground truth [194].

2.3.2 Compared Models

Ten state-of-the-art saliency models are evaluated for comparison. These models are either frequently cited in literature or have leading performance on the benchmark datasets. The compared models include spectral domain models (SigSal [74], HFT [107] and ΔQDCT [153]), models based on image patch statistics (LG [13] and AIM [19]), filter bank based methods (AWS [61] and Itti [79]), a learning based method (Judd [87]), a graph based method (GBVS [71]), and a context based method (CAS [64]). A summary of these models is provided in Table 2.2. Most of the models require off-line training or multi-scale processing. The initial version of BMS [201], denoted as BMS* is also included in the evaluation.

The code for these baseline methods is available on authors' websites,[2] and we use the default configurations set by the authors. The input image size for each model is listed in Table 2.2. Note that Judd's model [87] uses object detectors to facilitate saliency detection.

Table 2.2 Compared models

	Image size	Training[a]	Multi-scale Proc.
ΔQDCT[153]	64×48	No	Yes
SigSal[75]	64×48	No	No
LG[13]	512×512	Yes	Yes
AWS[61]	$\frac{1}{2}$ full size	No	Yes
HFT[107]	128×128	No	Yes
CAS[64]	$\max\{W, H\} = 250$	No	Yes
Judd[87]	200×200	Yes	Yes
AIM[19]	$\frac{1}{2}$ full size	Yes	No
GBVS[71]	Full size	No	Yes
Itti[79]	Full size	No	Yes
BMS*[201]	$W = 600$	No	No
BMS	$\max\{W, H\} = 400$	No	No

[a]Training includes dictionary learning

---

[2]For Itti's model, we use the improved version by [71].

2.3.3 Evaluation Methods

One of the most widely used metrics for saliency method evaluation is the ROC area under the curve (AUC) metric. Given a saliency map, we can generate a set of binary maps by thresholding. Each binary map represents a binary classification of image pixels, where positive pixels are predictions of eye fixations. Then the *true positive rate* (TPR) and *false positive rate* (FPR) can be computed based on a ground truth fixation map. By varying the threshold for generating the binary image, we can plot a curve of TPR against FPR. Then the AUC score is computed as the area under this curve. An advantage of this metric is that it only depends on the order of pixels rather than their absolute saliency values. Therefore, the AUC score is invariant to re-parametrization of a saliency map, as long as the pixel ordering is preserved [14].

However, factors such as border cut and center-bias setting have been shown to have a dramatic influence over the AUC metric[174, 205]. For example, in [205], it has been shown that a static Gaussian blob has an average ROC score of 0.80 on the Toronto dataset, exceeding many state-of-the-art methods, without using any features in the images. This phenomenon is due to the center bias of the spatial distribution of eye fixations on test images. Some methods explicitly or implicitly take advantage of the center bias, while others do not, which poses a challenge for fair comparisons. Note that although this prior spatial information is useful for eye fixation predictions when people view a still image on a screen, in general scenarios, when visual signals are collected in a more natural way (think of a robot navigating a room or a Google street view car collecting data), such center bias may not exist.

To control for the center bias effect in the benchmark datasets, a shuffled-AUC (sAUC) metric is proposed by [174, 205], which has become a standard evaluation method used in many recent works [13, 16, 61, 74, 153]. The sAUC is computed in a very similar way as the AUC. The only difference lies in the computation of FPR. For computing the sAUC score, we first need to compute a shuffle map $\mathcal{W}$ where for each pixel x, $\mathcal{W}(x)$ is the number of times that a fixation is located at x in the given dataset. For dataset composed of images with different resolutions, all the fixation maps need to be resized to a standard resolution before they are added up. Given an image and its ground truth map $\mathcal{C}$, the FPR is calculated based on a negative pixel set $\mathcal{N}$, which is sampled by the prior eye fixation distribution $P(x)$:

$$\text{FPR} = \frac{\#\{x \in \mathcal{N} : x \text{ is predicted as a fixation}\}}{\#\mathcal{N}},$$

$$\mathcal{N} = \{x : x \sim P(x)\},$$

$$P(x) = \frac{\delta(\mathcal{C}(x) = 0)\mathcal{W}(x)}{\sum_y \delta(\mathcal{C}(y) = 0)\mathcal{W}(y)}. \tag{2.10}$$

Note that negative sample set $\mathcal{N}$ may contain duplicate samples. Under the sAUC metric, common fixation positions will receive less credit for correct prediction. A perfect prediction will give an sAUC score of 1.0, while any fixed saliency map will give a score of approximately 0.5 [16]. An implementation of the shuffled-AUC metric is provided on our website.

Blurring can also significantly affect the sAUC scores. To control this factor, saliency maps are blurred with varying STD, and the mean sAUC scores on each dataset under optimal blurring are used to rank the models.

We also report the normalized scanpath saliency (NSS) scores [16, 135], which can provide a complementary view for the evaluation. The NSS score is computed as the mean saliency value of the fixation locations on a normalized saliency map, which has a zero mean and a unit standard deviation. We use an implementation of NSS provided by [86]. Similar to [183], we use a distance-to-center (DTC) re-weighting scheme to control the effect of center bias for fair comparison. A normalized center-to-distance map $\mathcal{J}$ is computed as follows:

$$\mathcal{J}(i, j) = 1 - \frac{\sqrt{\left(i - \frac{H}{2}\right)^2 + \left(j - \frac{W}{2}\right)^2}}{\sqrt{\left(\frac{H}{2}\right)^2 + \left(\frac{W}{2}\right)^2}}, \tag{2.11}$$

where i and j are the row index and the column index, respectively. To compute the NSS score of a saliency map, the saliency map is first smoothed by a Gaussian kernel with varying width, and then it is pixel-wise multiplied with $\mathcal{J}$ if this operation improves the score.

2.3.4 Results

For each model, the mean sAUC scores and mean NSS scores under the optimal postprocessing are presented in Table 2.3. The best scores are shown in red for each dataset, and the second and third best scores are underlined. The standard deviation of the sAUC scores ranges between 0.0005 and 0.001. We also report the *Human Inter-Observer* (IO) scores in the last column of the table. The IO scores are computed by randomly splitting the set of subjects in half, and using the fixations of one half as the ground truth, and the fixations of the other half as the saliency map, on which optimal blurring is also applied. The IO score represents the inter-observer consistency of a dataset, and serves as a performance upper bound for saliency models.

BMS consistently achieves the best sAUC and NSS scores on all the datasets except on the Cerf dataset. On the Cerf dataset, Judd's model ranks first and BMS ranks second under sAUC. However, the sAUC scores of BMS and Judd's model on the Cerf dataset are very close. Note that object detectors, including a face detector, are employed in Judd's model, and most of the test images in the

Table 2.3 Evaluation scores

Dataset	BMS	BMS* [201]	ΔQDCT [153]	SigSal [74]	LG [13]	AWS [61]	HFT [107]	CAS [64]	Judd [87]	AIM [19]	GBVS [71]	Itti [79]	Human IO
Mean sAUC with optimal blurring[a]													
MIT	0.7105	0.6932	0.6741	0.6685	0.6813	0.6939	0.6514	0.6719	0.6729	0.6706	0.6351	0.6440	0.7756
Opt. σ	0.04	0.05	0.04	0.04	0.06	0.00	0.00	0.05	0.04	0.05	0.00	0.05	0.06
Toronto	0.7243	0.7206	0.7168	0.7054	0.6986	0.7174	0.6900	0.6955	0.6908	0.6906	0.6391	0.6576	0.7316
Opt. σ	0.05	0.03	0.00	0.00	0.05	0.00	0.02	0.04	0.05	0.04	0.00	0.02	0.07
Kootstra	0.6276	0.6207	0.6014	0.6000	0.6053	0.6231	0.5874	0.6011	0.5940	0.5906	0.5543	0.5780	0.6854
Opt. σ	0.03	0.00	0.01	0.00	0.04	0.00	0.02	0.03	0.04	0.00	0.00	0.00	0.05
Cerf	0.7491	0.7357	0.7281	0.7282	0.7028	0.7247	0.6994	0.7141	0.7517	0.7243	0.6800	0.6764	0.7944
Opt. σ	0.01	0.02	0.04	0.02	0.06	0.00	0.03	0.04	0.04	0.04	0.01	0.06	0.07
ImgSal	0.7078	0.7030	0.6835	0.6753	0.6755	0.6990	0.6779	0.6856	0.6778	0.6703	0.6432	0.6491	0.7411
Opt. σ	0.05	0.05	0.05	0.00	0.06	0.04	0.03	0.05	0.05	0.04	0.00	0.04	0.06
SBU-VOC	0.6876	0.6790	0.6652	0.6581	0.6550	0.6675	0.6404	0.6659	0.6644	0.6544	0.6022	0.6162	0.6946
Opt. σ	0.08	0.08	0.10	0.08	0.10	0.07	0.06	0.08	0.09	0.07	0.03	0.09	0.11
DUT-O	0.7365	0.7258	0.7091	0.7077	0.6983	0.7255	0.6958	0.7204	0.7188	0.6999	0.6556	0.6665	0.7694
Opt. σ	0.05	0.06	0.07	0.05	0.06	0.04	0.05	0.07	0.05	0.05	0.01	0.03	0.07
Avg.	0.7062	0.6969	0.6826	0.6776	0.6738	0.6930	0.6632	0.6792	0.6815	0.6715	0.6299	0.6411	0.7417

Dataset	BMS	BMS*	*QDCT	SigSal	LG	AWS	HFT	CAS	Judd	AIM	GBVS	Itti	Human
		[201]	[153]	[74]	[13]	[61]	[107]	[64]	[87]	[19]	[71]	[79]	IO
Mean NSS with optimal blur σ and distance-to-center re-weighting													
MIT	1.742	1.538	1.535	1.513	1.509	1.589	1.439	1.508	1.450	1.327	1.534	1.437	2.708
Opt σ	0.03	0.04	0.05	0.05	0.05	0.01	0.07	0.05	0.00	0.04	0.01	0.07	0.05
Toronto	1.933	1.399	1.776	1.736	1.597	1.757	1.622	1.698	1.472	1.397	1.659	1.572	3.274
Opt σ	0.03	0.00	0.00	0.00	0.04	0.00	0.03	0.03	0.00	0.02	0.00	0.01	0.04
Koostra	0.8701	0.7190	0.7104	0.7058	0.7318	0.7789	0.6776	0.7110	0.6559[b]	0.6341	0.6611	0.6357	1.414
Opt σ	0.02	0.02	0.03	0.03	0.04	0.02	0.07	0.04	0.01	0.02	0.01	0.05	0.05
Cerf	1.626	1.529	1.500	1.477	1.321	1.439	1.404	1.418	1.373[b]	1.200	1.413	1.313	2.752
Opt σ	0.03	0.03	0.03	0.01	0.05	0.01	0.04	0.04	0.01	0.03	0.01	0.06	0.04
ImgSal	1.877	1.813	1.672	1.634	1.534	1.746	1.680	1.680	1.463	1.291	1.644	1.562	2.747
Opt σ	0.03	0.03	0.00	0.00	0.04	0.00	0.04	0.04	0.00	0.03	0.00	0.03	0.04
SBU-VOC	1.596	1.572	1.482	1.447	1.409	1.486	1.414	1.475	1.436	1.308	1.467	1.435	1.840
Opt σ	0.07	0.07	0.08	0.09	0.08	0.08	0.10	0.07	0.00	0.06	0.05	0.09	0.10
DUT-O	2.158	2.055	1.882	1.893	1.726	1.976	1.855	1.904	1.703	1.475	1.835	1.772	2.951
Opt σ	0.02	0.02	0.03	0.00	0.03	0.01	0.04	0.03	0.01	0.03	0.01	0.01	0.06
Avg	1.686	1.618	1.508	1.487	1.404	1.539	1.442	1.485	1.365	1.233	1.459	1.389	2.527

The best score on each dataset is shown in red. The second and third best are underlined

[a] The blur STD is represented as the ratio to the largest dimension of the saliency maps

[b] These AUC scores are computed without the distance-to-center re-weighting, as they are not improved by this post-processing

Cerf dataset contain human faces as the key salient objects. BMS also consistently improves over our previous version, denoted as BMS* [201], under sAUC and NSS.

Under the sAUC metric, the scores of the leading models are close to those of the IO baseline on the Toronto and SBU-VOC datasets, while under the NSS metric, the gap between IO and other models on the Toronto and SBU-VOC datasets is still substantial. This reflects the different characteristics of the sAUC and NSS metrics, and there is still notable difference between the performance of human and computational models on all the datasets. There is a low inter-observer consistency on the Kootstra dataset, as indicated by the IO scores. Consequently, the compared models perform much worse on this dataset than on the other ones.

Since blurring and distance-to-center re-weighting play a very important role in comparing different models, we further present some relevant statistics for analyzing the influence of these factors in evaluation. First, we show the influence of blurring on the sAUC scores in Fig. 2.4. On the MIT, SBU-VOC, and DUT-O datasets, BMS outperforms the competing models by a considerable margin over a wide range of blur levels. The test images in these three datasets are mostly composed of everyday pictures, which usually contain objects of medium and large scale. To handle salient regions of different sizes, many competing models resort to multi-scale processing. Due to the scale-invariant nature of the surroundedness cue, BMS can better handle these scenarios without using any multi-scale processing. Blurring has a relatively more drastic impact on the sAUC scores of AIM on the Toronto, Kootstra, Cerf, and ImgSal datasets, and on the scores of LG on the Cerf dataset. Otherwise the rankings of the models are quite consistent over a wide range of blur levels.

We also evaluate the influence of the distance-to-center (DTC) re-weighting over the NSS metric in Fig. 2.5. We control the blur factor and compare the NSS scores computed with DTC re-weighting and without. The NSS scores of almost all the models are significantly improved with DTC re-weighting. Compared with the other models, Judd and GBVS do not benefit much from this post-processing, and the NSS scores of Judd even decrease on the Kootstra and Cerf datasets with the DTC re-weighting. Judd's model explicitly employs an optimized distance-to-center feature map, and GBVS produces saliency maps with a strong center bias [107]. Therefore, the DTC re-weighting does not improve their scores as much as the other models. Without using DTC re-weighting to calibrate the NSS scores, models that take advantage of the center bias often rank higher than other ones.

By qualitatively examining the saliency maps, we found that BMS tends to be less distracted by high-contrast edge areas than most of the compared models, and it can better highlight the interior regions of salient objects of different sizes, even though it does not use multi-scale processing. Some sample images and saliency maps of compared models are shown in Fig. 2.6. Figure 2.6a–e shows the images with salient regions of different sizes. In (a) and (b), the salient regions are very small (see Fig. 2.7), and many compared models fail to detect them. In contrast,

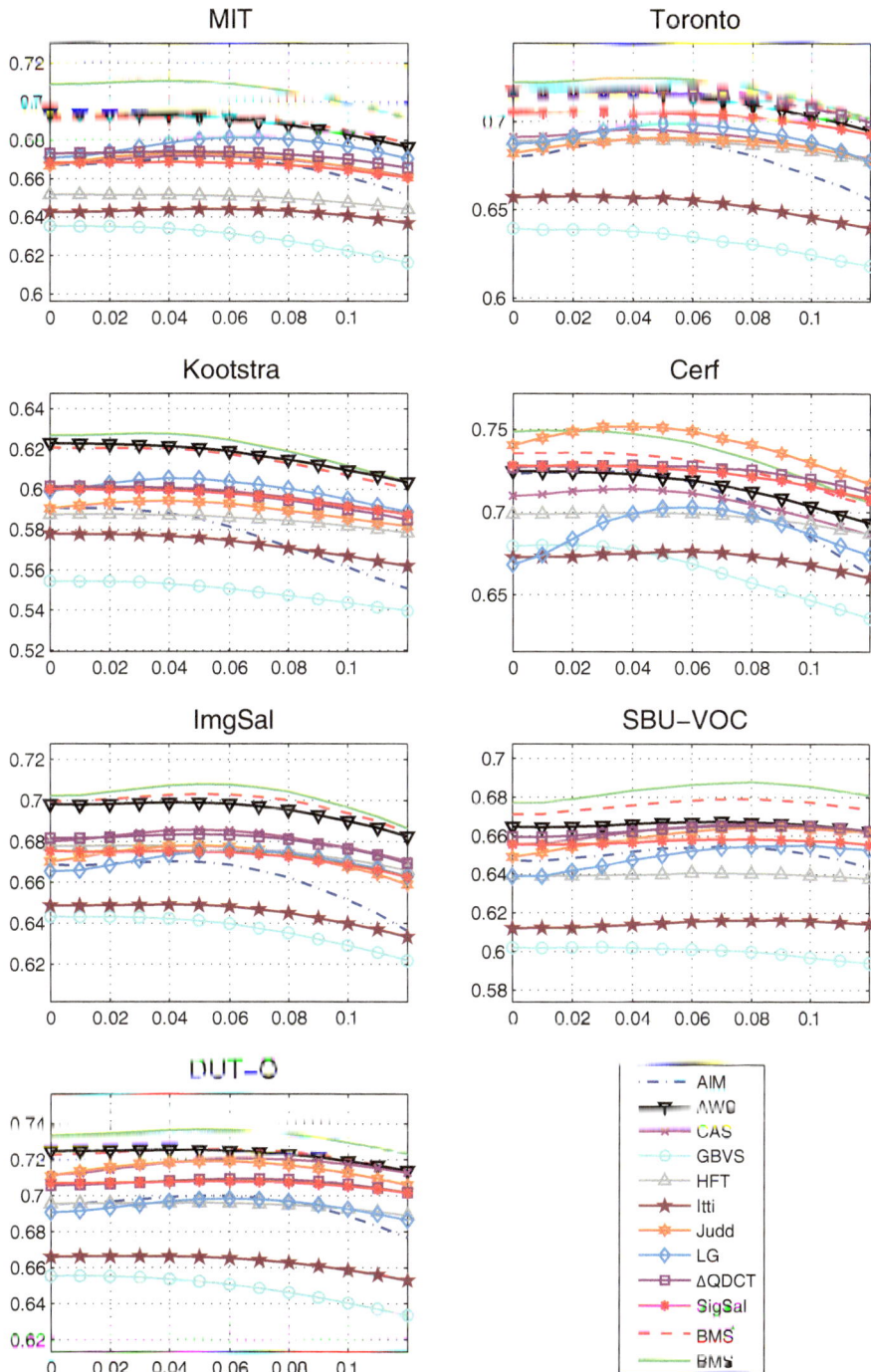

Fig. 2.4 sAUC scores against the level of blur. For each plot, the *x* axis is the Gaussian blur STD relative to the largest dimension of the saliency maps, and the *y* axis is the sAUC score

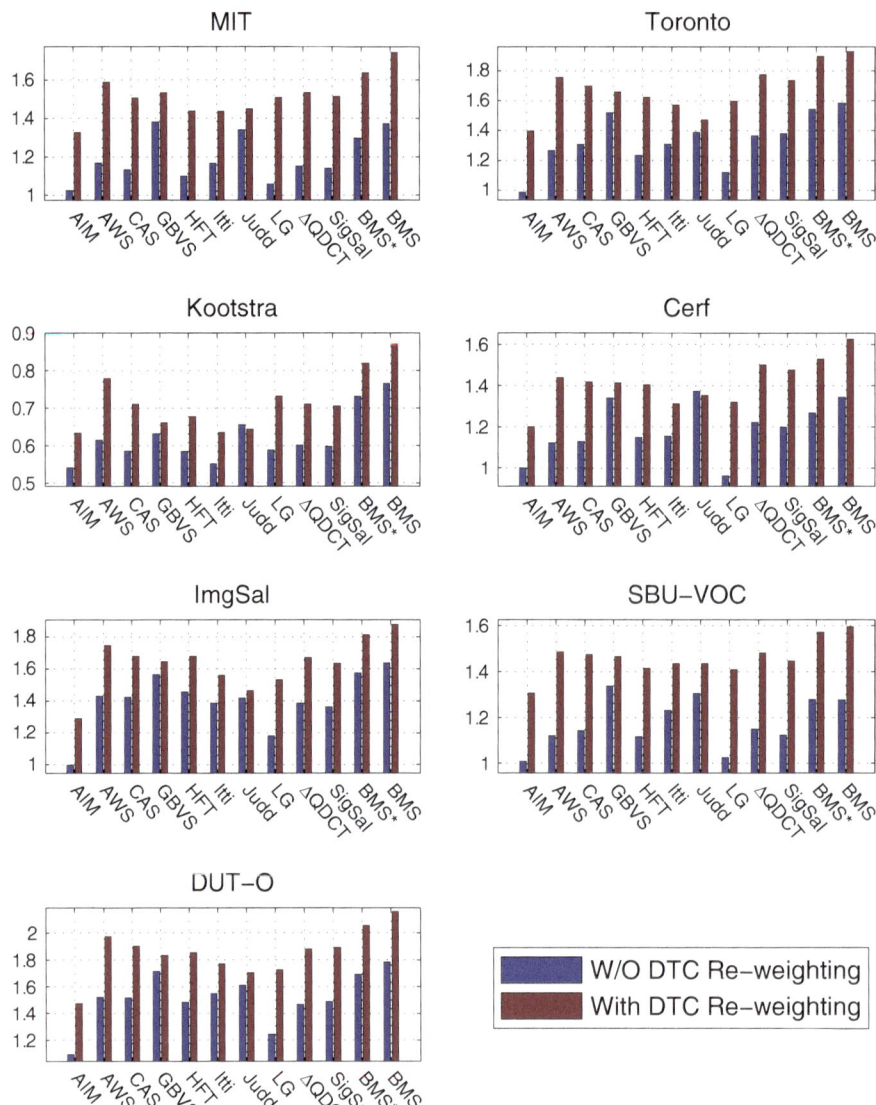

Fig. 2.5 The influence of distance-to-center re-weighting on the NSS scores. On each plot, the blue bars represent the NSS scores under the optimal blurring computed *without* distance-to-center (DTC) re-weighting, and the red bars represent the NSS scores under the optimal blurring computed *with* DTC re-weighting

BMS can accurately detect these small regions. In (c)–(e), BMS is less responsive to the object boundaries and edge areas in the background, leading to more accurate saliency maps. In Fig. 2.6f–i, sample images contain human faces of various sizes. BMS can more correctly highlight the interior area of faces than most of the

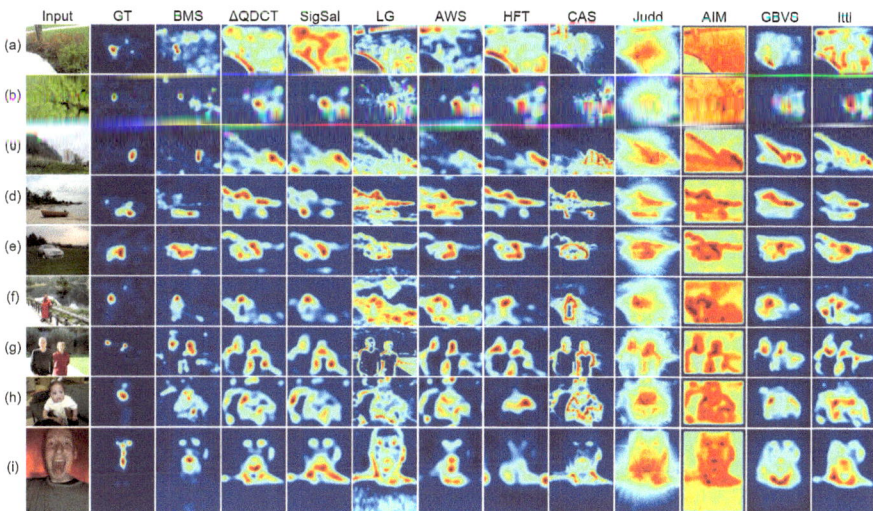

Fig. 2.6 Sample images and saliency maps of compared models. The second column shows the eye fixation heat maps, which are obtained by blurring the eye fixation maps

Fig. 2.7 A closer look at the input images in Fig. 2.6a, b

compared models, owing to the enclosure relationship between the face and the background (see (f) and (g)), and between the facial features (eyes and mouth) and the face (see (h) and (i)).

Speed Performance We compare the speed performance of all the models on a machine with a quad-core 2.93 GHz CPU and 8 GB memory using a single thread. We measure the speed of all the models on the Toronto dataset, where the size of the input images is 685×511. Note that the compared models may resize the input images in different ways (see Table 2.2). We exclude the time for reading and writing images, so that the speed measurement is more accurate, especially for the faster models. BMS is implemented in C, and all the other models are implemented in Matlab or Matlab+C. The speed performance in FPS is shown in Fig. 2.8. BMS runs at about 11 FPS, which is the second fastest model. BMS obtains an approximately 4X speedup compared with our previous version BMS*, which runs at about three FPS. Half of the speedup is attributed to the smaller image size used by BMS. The rest of the speedup is attributed to modification of the activation map computation (see Sect. 2.2.2).

Fig. 2.8 Speed performance comparison

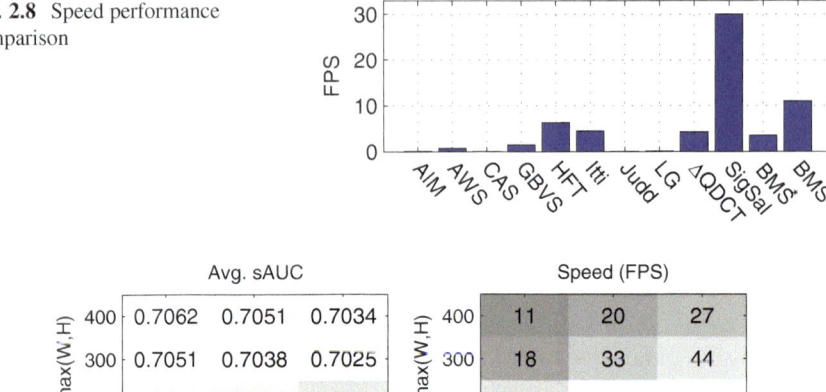

Fig. 2.9 Speed and performance of BMS under different configurations. The left plot shows the average mean sAUC scores of BMS over all the datasets with different settings for the input image size and the threshold sample step. The right plot shows the corresponding speed in FPS for each configuration. Whiter cells indicate larger values

Spectral domain based models such as SigSal, ΔQDCT, HFT, are fast, because they usually down-sample the input image to a very small size. Models that involve multi-scale convolutions or patch based dictionaries are relatively slower, e.g. AIM, AWS, and LG. For Judd's model, most of the time is spent on object detections, which can take up to 10 s. CAS requires an exhaustive computation of all pairs of segments, and it takes about 30 s to process an image.

2.3.5   Speed Accuracy Tradeoff

The speed of BMS depends on two factors: the input image size and the threshold sample step. We show in Fig. 2.9 how we can trade off a little performance for a remarkable gain in speed. We vary the maximum dimension of input images among 200, 300, and 400 pixels, and the sample step among 8, 16, and 24, assuming an 8-bit three-channel image. Note that in this test, all the kernel widths for the blurring and dilation operations scale linearly to the maximum dimension of the input image. The performance of BMS is measured by its average mean sAUC scores under the optimal blurring over all the datasets. As shown in Fig. 2.9, with a little drop in performance, BMS can be remarkably accelerated. For example, BMS has a 4X speedup by setting the maximum image dimension to 300 and the sample step to 24, and its average mean sAUC is 0.7025, only slightly lower than the score of the original configuration. BMS can even run at about 100 FPS, with an average mean sAUC score of 0.6983, which still compares favorably with the average mean sAUC scores of other competing models (see Table 2.3). This flexible characteristic of BMS makes it suitable for many time-critical applications.

2.3.6   Component Analysis

First, we analyze the effect of color space whitening. We show in Fig. 2.10 the mean sAUC scores of BMS with and without color space whitening under the RGB, CIE LUV, and CIE Lab color spaces. Without color space whitening, the RGB space is substantially inferior to the LUV and the Lab space. Note that the LUV and the Lab space are known for their perceptual uniformity. After the color space whitening, the scores from different color spaces become very close. Overall, color space whitening is beneficial for BMS using all of the three color spaces, which validates our analysis of Boolean map sampling in Sect. 2.2.1.

Next, we compare the performance of BMS with and without splitting the activation map before normalization in Fig. 2.11. We see a consistent performance gain with activation map splitting across all the datasets. In our previous version, BMS*, the activation map is directly normalized without splitting. Activation map

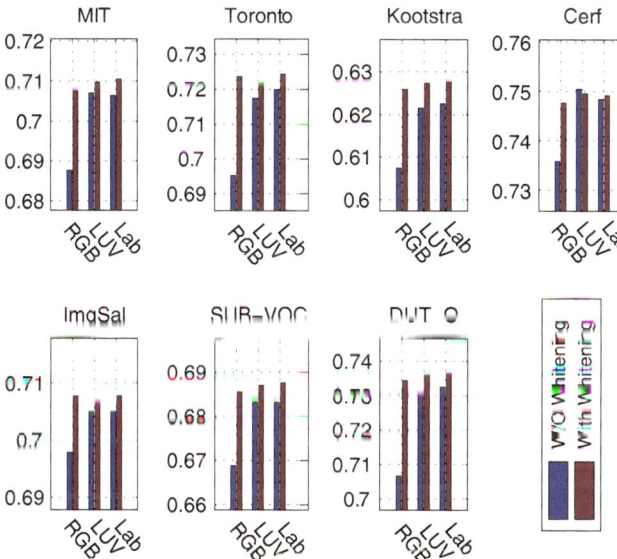

Fig. 2.10 Mean sAUC scores of BMS with and without color space whitening using the RGB, CIE LUV, and CIE Lab color spaces

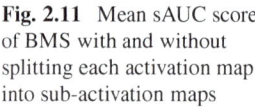

Fig. 2.11 Mean sAUC score of BMS with and without splitting each activation map into sub-activation maps

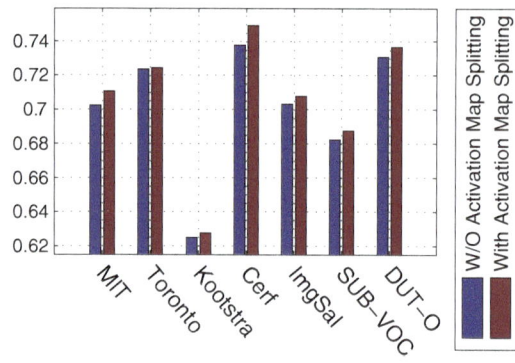

Fig. 2.12 The effects of the dilation kernel width ω and κ over the mean sAUC scores of BMS on each dataset

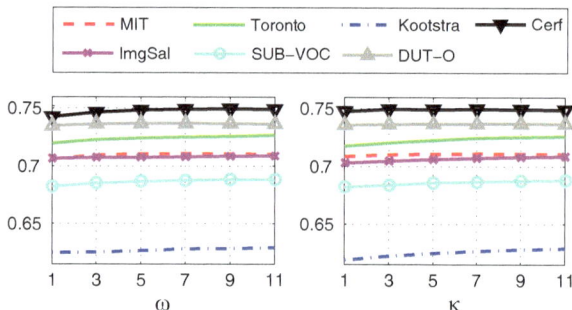

splitting, together with the feature space whitening, contributes to the overall performance improvement of BMS over its previous version.

Finally, we give a parameter analysis of BMS. The effects of blurring, input image size, and threshold sample step are already discussed. Therefore, there are only two remaining parameters, the dilation kernel width ω involved in the normalization step and the dilation kernel width κ involved in the postprocessing step. In Fig. 2.12 the changes of sAUC scores of BMS by varying these two parameters on each dataset are shown. These two dilation operations generally improve the performance of BMS. However, the performance of BMS is not very sensitive to the settings of these two parameters.

2.4 Conclusion

In this chapter, we proposed a Boolean map based saliency model, which exploits the surroundedness cue for eye fixation prediction. In BMS, an image is first decomposed into a set of Boolean maps by randomly thresholding the color channels, and then the surrounded regions on each Boolean maps are activated. The resultant activation maps are normalized and linearly combined to generate a saliency map.

In experiments, despite its simplicity, the BMS model consistently outperforms ten state-of-the-art models on seven datasets. BMS is also capable of running at about 100 FPS with competitive performance, which makes it suitable for time-critical applications.

Our proposed BMS model demonstrates the effectiveness of the surroundedness cue and provides a very efficient way to leverage it for eye fixation prediction. The surroundedness cue captures the global topological feature for holistic figure-ground segmentation, which is one of the important Gestalt elements that affect human visual attention [133]. However, previous works [13, 19, 79] were mainly based on the contrast/rarity cue and spectral analysis, and thus could not model the holistic figure-ground segmentation phenomenon.

It is worth noting that previous eye fixation prediction models are often quite sensitive to the choice of the *scale* parameters, e.g. the scale of the input image, the scale of the image patch descriptor, and the scale of the image filter. Without properly selecting a suitable scale, previous methods tend to either overemphasize the boundaries of large regions or miss small salient regions (see Fig. 2.6). In contrast, the surroundedness cue is quite robust to the scale issue. As a result, our BMS model can detect salient regions of very different scales without using multi-scale processing. This adds to the efficiency of our BMS model.

Due to the superior efficiency and quality of our BMS model, various applications have been developed based on our BMS model, e.g. automatic eye tracker calibration [168], glottis segmentation from videos [154], image compression [103], low-bit coding for video chat [197], object recognition [117], image profiling [28], and maritime surveillance [163].

While the surroundedness cue is a quite general assumption about salient regions, our BMS model still needs the other cues, e.g. the contrast cue, to enhance its performance. In BMS, we employ additional techniques such as normalization and activation map splitting to better detect rare regions. This indicates that using a single cue may not be sufficient, and the combination of different and potentially complementary features should be the key to advancing the state of the art. In particular, simple cues such as surroundedness and contrast cannot capture high-level saliency features such as objects and faces. Therefore, in scenarios where human visual attention is mostly influenced by high-level features, our BMS model tends to fail.

Chapter 3
A Distance Transform Perspective

Distance functions and their transforms (DTs, where each pixel is assigned the distance to a set of *seed pixels*) are used extensively in many image processing applications. In this chapter, we will provide a distance transform perspective for the core algorithm of BMS. We show that the core algorithm of BMS is basically an efficient distance transform algorithm for a novel distance function, the *Boolean map distance* (BMD). We will further link the Boolean map distance with the *minimum barrier distance* (MBD), which was first proposed by Strand et al. [166], and show that the BMD shares many of the favorable properties of the MBD, while offering some additional advantages such as more efficient distance transform computation and straightforward extension to multi-channel images. This connection between the BMD and the MBD also leads to alternative algorithms to approximate the BMD and MBD transforms.

3.1 The Boolean Map Distance

In this section, we introduce the Boolean map distance (BMD) in the discrete image space. The corresponding formulation in the continuous image domain can be found in [120]. To formally define the Boolean map distance, we need a few basic concepts in the discrete image domain.

Grayscale Image We define a discrete grayscale digital image $\mathcal{I}$ as a pair $\mathcal{I} = (V, \mathcal{F})$ consisting of a finite set V of image elements and a mapping $\mathcal{F} : V \to [0, 1]$.[1] We will refer to elements of V as *pixels*, regardless of the dimensionality of the image. Additionally, we define a mapping $\mathcal{N} : V \to \mathcal{P}(V)$ specifying an

---

[1] The restriction of the image values to the range [0, 1] does, for the purposes considered here, not imply a loss of generality.

© Springer Nature Switzerland AG 2019
J. Zhang et al., *Visual Saliency: From Pixel-Level to Object-Level Analysis*,
https://doi.org/10.1007/978-3-030-04831-0_3

adjacency relation over the set of pixels V. For any $s, t \in V$, we refer to $\mathcal{N}(s)$ as the *neighborhood* of s and say that t is *adjacent* to s if $t \in \mathcal{N}(s)$. We require the adjacency relation to be symmetric, so that $t \in \mathcal{N}(s) \Leftrightarrow s \in \mathcal{N}(t)$ for all $s, t \in V$.

Path   A discrete path $\pi = \langle \pi(0), \pi(1), \ldots \pi(L) \rangle$ of length $|\pi| = L + 1$ from $\pi(0)$ to $\pi(L)$ is an ordered sequence of pixels in V where each consecutive pair of pixels is *adjacent*. We use the symbol $\Pi_{s,t}^V$ to denote the set of all discrete paths from s to t. For a set of pixels $S \subseteq V$, the symbol $\Pi_{S,t}^V$ denotes the set of all discrete paths from any $s \in S$ to t. The *reverse* π^{-1} of a discrete path π is defined as $\pi^{-1}(i) = \pi(L - i)$ for all $i \in \{0, 1, \ldots, L\}$. Given two discrete paths π_1 and π_2 such that the endpoint of π_1 equals the starting point of π_2, we denote by $\pi_1 \cdot \pi_2$ the concatenation of the two paths.

Path Connected V is *path connected* if for every $s, t \in V$ there exists a path π from s to t. For the remainder of this section, we assume that the set V is path connected.

For simplicity, we further assume that V is a regular path-connected pixel space, and $\mathcal{F}$ and $\mathcal{I}$ are used interchangeably to denote an image.

Boolean Map   We refer to any function that maps image elements to the set $\{0, 1\}$ as a *Boolean map*. Here we define a Boolean map function based on image thresholding. Let $k \in [0, 1]$. Given an image $\mathcal{F}$, we define the function $\mathcal{B}_k^{\mathcal{F}} : D \to \{0, 1\}$ by

$$\mathcal{B}_k^{\mathcal{F}}(s) = \begin{cases} 0, & \text{if } \mathcal{F}(s) < k \\ 1, & \text{otherwise} \end{cases}, \tag{3.1}$$

The Boolean map $\mathcal{B}_k^{\mathcal{F}}$ represents the *thresholding* of the image $\mathcal{F}$ by k.

Connected Component   For any $s, t \in V$, we say that s and t belong to the same component of $\mathcal{B}_k^{\mathcal{F}}$ if there exists a path $\pi \in \Pi_{s,t}^V$ such that either $\mathcal{B}_k^{\mathcal{F}}(\pi(i)) = 0$ for all $i \in \{0, 1, \ldots, L\}$ or $\mathcal{B}_k^{\mathcal{F}}(\pi(i)) = 1$ for all $i \in \{0, 1, \ldots, L\}$. A path satisfying either of these criteria is called a *connecting path*. We use the notation $s \underset{k}{\sim} t$ to indicate that s and t belong to the same component of $\mathcal{B}_k^{\mathcal{F}}$, while the notation $s \underset{k}{\nsim} t$ indicates that they belong to different components. Additionally, for a set of pixels S, the notation $S \underset{k}{\sim} t$ indicates that $s \underset{k}{\sim} t$ for at least one $s \in S$ while the notation $S \underset{k}{\nsim} t$ indicates that $s \underset{k}{\nsim} t$ for all $s \in S$.

Now we introduce the definition of the Boolean map distance and show some of its properties.

Definition 3.1   Let t be a random value sampled from a uniform probability distribution over $[0, 1]$. The *Boolean map distance* on a discrete image BMD: $\mathcal{P}(V) \times V \to [0, 1]$ is defined as

$$\text{BMD}(S, t) = P(S \underset{k}{\nsim} t) = 1 - P(S \underset{k}{\sim} t) \tag{3.2}$$

In the above definition, $\mathcal{P}(V)$ denotes the power set of V. If the set S consists of a single element s, we can consider BMD to be a mapping from $V \times V$ to $[0, 1]$, and the definition can in this case be reduced to $\mathrm{BMD}(s, t) = P(s \underset{k}{\approx} t)$.

Definition 3.2 A function $d : V \times V \to [0, \infty)$ is a *pseudo-metric* on a set V if, for every $s, t, p \in V$,

1. $d(s, s) = 0$ (identity)
2. $d(s, t) \geq 0$ (non-negativity)
3. $d(s, t) = d(t, s)$ (symmetry)
4. $d(s, p) \leq d(s, t) + d(t, p)$ (triangle inequality)

If additionally it holds that $d(p, q) = 0 \Leftrightarrow p = q$ for all p, q, then d is a *metric*.

Theorem 3.3 *Let $S \subseteq V$ consist of a single element s and let $t \in V$. Then the Boolean map distance, viewed as a mapping from $V \times V$ to $[0, 1]$, is a pseudo-metric.*

Proof First, we show that BMD obeys property (i). Consider a path π such that it only contains the pixel s. This path is obviously connecting s to itself. Thus, $s \underset{k}{\sim} s$ for all k, and so $P(s \underset{k}{\approx} s) = 0$.

Since BMD is defined as a probability, we have $\mathrm{BMD}(s, t) \in [0, 1]$ for all s, t. Thus, BMD clearly obeys property (ii).

Next, we show that BMD obeys property (iii). If, for a given threshold k, there exists a connecting path π from s to t, then the reverse path π^{-1} is a connecting path from t to s. Thus, $s \underset{k}{\sim} t \Leftrightarrow t \underset{k}{\sim} s$, and so $P(s \underset{k}{\approx} t) = P(t \underset{k}{\approx} s)$.

Finally, we show that BMD obeys property (iv). If, for a given threshold k, there exists a connecting path π_1 from s to t and another connecting path π_2 from t to p, then the concatenation $\pi_1 \cdot \pi_2$ of these two paths is a connecting path from s to p. Thus, the set of thresholds k for which $s \underset{k}{\sim} t$ and $t \underset{k}{\sim} p$ is a subset of the set of thresholds for which $s \underset{k}{\sim} p$, and so $\mathrm{BMD}(s, p) \leq \mathrm{BMD}(s, t) + \mathrm{BMD}(t, p)$. □

Note that for a constant function $\mathcal{F}$ we have $P(s \underset{k}{\approx} t) = 0$ for all $s, t \in D$, and thus BMD is not in general a metric.

3.2 BMS and the Boolean Map Distance Transform

Given a discrete image $\mathcal{I} = (V, \mathcal{F})$ and a set of *seed pixels* $S \subseteq V$, the *distance transform* for the Boolean map distance (BMD) is a map assigning to each pixel $t \in V$ the value $\mathrm{BMD}(S, t)$, i.e., each pixel is assigned the Boolean map distance to the set S.

From the definition of the BMD, it is straightforward to devise a Monte Carlo algorithm for approximating the BMD distance transform by iteratively selecting a

random threshold, performing thresholding, and using a flood-fill operation to find
the set of pixels connected to at least one seed-point. This is basically the same as
the activation map computation in Boolean map saliency (BMS).

Assume that all intensities present in a given image can be written as i/N, for
some fixed integer N and some i in the set $\{0, 1, \ldots, N-1\}$. This situation occurs in
practice if we, e.g., remap an image with integer intensity values to the range $[0, 1]$.
If gray levels are stored as 8-bit integers, for example, we can take $N = 256$. Then
the algorithm for calculating the Boolean map saliency (see the main for-loop of
Algorithm 1 in Chap. 2) can be directly used for computing the exact BMD distance
transform from any set of seed-points. Pseudo-code for this algorithm is listed in
Algorithm 1.

In this algorithm, each iteration of the **foreach**-loop requires $\mathcal{O}(n)$ operations,
where n is the number of image pixels. Thus, the entire algorithm terminates
in $\mathcal{O}(nN)$ operations which, since N can be considered a constant, equals $\mathcal{O}(n)$
operations.

Algorithm 1 A Monte Carlo BMD Distance Transform Algorithm

Input: An image $\mathcal{I}$, a set of seed-points S, an integer N
Output: Distance transform T
1: Set $T(s) = 0$ for all pixels s in $\mathcal{I}$;
2: **for** $i \in \{0, 1, \ldots, N - 1\}$ **do**
3:     Set $B \leftarrow \mathcal{B}^{\mathcal{F}}_{i/N}$;
4:     Perform a flood-fill operation to identify the set of pixels belonging to the same component
        as at least one seed-point in B;
5:     Increase T by $1/N$ for all pixels not in this set;
6: **end for**

3.3 BMS and the Minimum Barrier Distance Transform

Now we show a connection between BMS and the minimum barrier distance (MBD)
transform [40, 166], which reveals the relationship between BMD and the MBD.
Compared with other distances, e.g. geodesic distance and fuzzy distance, MBD is
advantageous for its robustness to blurring and noise. These nice properties of MBD
shed light on why BMS or BMD can properly capture the surroundedness cue in an
image.

3.3.1 Preliminaries

On a grayscale image $\mathcal{I} = (V, \mathcal{F})$, we define the following two path functions:

$$\beta^{+}_{\mathcal{F}}(\pi) = \max_i \mathcal{F}(\pi(i)), \tag{3.3}$$

$$\beta^{-}_{\mathcal{F}}(\pi) = \min_i \mathcal{F}(\pi(i)), \tag{3.4}$$

where $\pi = \{\pi(i)\}_{i=0}^{L}$ is a path. Think of $\mathcal{F}$ as an elevation map, and then $\beta_{\mathcal{F}}^{+}(\pi)$ (*resp.* $\beta_{\mathcal{F}}^{-}(\pi)$) represents the height of the highest (*resp.* lowest) point on a path.

Given a seed set S and a pixel t, let $\Pi_{S,t}$ denote the set of paths joining an element of S and t. The *minimum barrier distance* (MBD) [40, 166] is defined as

$$d_{\mathcal{F}}(S, t) = \min_{\pi \in \Pi_{S,t}} \left(\beta_{\mathcal{F}}^{+}(\pi) - \beta_{\mathcal{F}}^{-}(\pi) \right). \tag{3.5}$$

In the above formulation, the length of a path π is defined as the elevation from its lowest point to its highest point, *a.k.a.* the *barrier* of π. The MBD between S and t is the length of the shortest path in $\Pi_{S,t}$.

It is straightforward to see that

$$d_{\mathcal{F}}(S, t) = \min_{\pi \in \Pi_{S,t}} \left(\beta_{\mathcal{F}}^{+}(\pi) - \beta_{\mathcal{F}}^{-}(\pi) \right),$$

$$\geq \min_{\pi \in \Pi_{S,t}} \beta_{\mathcal{F}}^{+}(\pi) - \max_{\pi' \in \Pi_{S,t}} \beta_{\mathcal{F}}^{-}(\pi').$$

Let

$$\varphi_{\mathcal{F}}(S, t) := \min_{\pi \in \Pi_{S,t}} \beta_{\mathcal{F}}^{+}(\pi) - \max_{\pi' \in \Pi_{S,t}} \beta_{\mathcal{F}}^{-}(\pi'). \tag{3.6}$$

Then $\varphi_{\mathcal{F}}(S, t)$ is a lower bound of the MBD $d_{\mathcal{F}}(S, t)$, and it can be regarded as another distance function *w.r.t.* S and t. Note that $\varphi_{\mathcal{F}}(S, t) \geq 0$, because $\beta_{\mathcal{F}}^{+}(\pi) \geq \mathcal{F}(t)$ and $\beta_{\mathcal{F}}^{-}(\pi) \leq \mathcal{F}(t)$ for all $\pi \in \Pi_{S,t}$.

Given a distance function f (e.g. $\varphi_{\mathcal{F}}$ or $d_{\mathcal{F}}$) and a seed set S, we denote a *distance transform* as Θ_{f}^{S},

$$\Theta_{f}^{S}(t) = f(S, t). \tag{3.7}$$

As the distance function, either $d_{\mathcal{F}}$ or $\varphi_{\mathcal{F}}$, is dependent on the image $\mathcal{F}$, we can think of the distance transform as a function of the image. Therefore, the following notations are introduced:

$$\Theta_{\varphi}^{S}(\mathcal{F}) := \Theta_{\varphi_{\mathcal{F}}}^{S}, \tag{3.8}$$

$$\Theta_{d}^{S}(\mathcal{F}) := \Theta_{d_{\mathcal{F}}}^{S}. \tag{3.9}$$

3.3.2   BMS and $\varphi_{\mathcal{F}}$

As discussed before, a real-valued image can always be discretized by shifting, scaling, and rounding its values with a desirable precision, so that the values of the discrete image are in a N-level space $\{0, 1, \cdots N - 1\}$. Thus, we assume each feature map of an image is an N-level image. On such a feature map $\mathcal{F}$, if the threshold sample step $\delta = 1$, then the set of the generated Boolean maps will be

$$\Gamma(\mathcal{F}) := \{\mathcal{B}_k^{\mathcal{F}}\}_{k=0}^{N-1},$$

$\Gamma(\mathcal{F})$ forms a *threshold decomposition* [122, 184] of $\mathcal{F}$. It is easy to see that $\mathcal{F}$ can be reconstructed from $\Gamma(\mathcal{F})$ via $\mathcal{F} = \sum_k \mathcal{B}_k^{\mathcal{F}}$.

The concept of the threshold decomposition can be used to link a binary image transform to a grayscale image transform. Let Ψ denote a transform defined on grayscale images. Ψ obeys the *linear-threshold superposition* [122] if the following holds:

$$\Psi(\mathcal{F}) = \sum_{k=0}^{N-1} \Psi(\mathcal{B}_k^{\mathcal{F}}). \tag{3.10}$$

In what follows, we show the distance transform $\Theta_\varphi^S(\mathcal{F})$ induced by Eq. 3.6 obeys the linear-threshold superposition.

Lemma 3.4 *Given an N-level image $\mathcal{I} = (V, \mathcal{F})$ and a seed set S, the distance transform $\Theta_\varphi^S(\mathcal{F})$ obeys the linear-threshold superposition:*

$$\Theta_\varphi^S(\mathcal{F}) = \sum_{k=0}^{N-1} \Theta_\varphi^S(\mathcal{B}_k^{\mathcal{F}}). \tag{3.11}$$

Proof For convenience, let

$$h_{\mathcal{F}}^+(S, t) \triangleq \min_{\pi \in \Pi_{S,t}} \beta_{\mathcal{F}}^+(\pi), \tag{3.12}$$

$$h_{\mathcal{F}}^-(S, t) \triangleq \max_{\pi' \in \Pi_{S,t}} \beta_{\mathcal{F}}^-(\pi'), \tag{3.13}$$

$$\beta_{\mathcal{F}}(\pi) \triangleq \beta_{\mathcal{F}}(\pi)^+ - \beta_{\mathcal{F}}(\pi)^-. \tag{3.14}$$

Therefore, $\varphi_{\mathcal{F}}(S, t) = h_{\mathcal{F}}^+(S, t) - h_{\mathcal{F}}^-(S, t)$ and $d_{\mathcal{F}}(S, t) = \min_{\pi \in \Pi_{S,t}} \beta_{\mathcal{F}}(\pi)$. To avoid unnecessary clutter in notation, we use Θ to denote Θ_φ^S, and $\mathcal{B}_k$ to denote $\mathcal{B}_k^{\mathcal{F}}$.

Recall that for each pixel t,

$$\Theta(\mathcal{B}_k)(t) = h_{\mathcal{B}_k}^+(S, t) - h_{\mathcal{B}_k}^-(S, t). \tag{3.15}$$

When $\mathcal{B}_k(t) = 1$, we have $h_{\mathcal{B}_k}^+(S, t) = 1$. In this case, $\Theta(\mathcal{B}_k)(t) = 0$ iff there is a path π joining a seed $s \in S$ and t, such that $\beta_{\mathcal{B}_k}^-(\pi) = 1$. This means that π

is in the white set of $\mathcal{B}_k$, and it follows that $\beta_{\bar{\mathcal{F}}}(\pi) > k$. Thus, when $\mathcal{B}_k(t) = 1$, $\Theta(\mathcal{B}_k)(t) = 0$ iff there exists a path π joining a seed s and t, where $\beta_{\bar{\mathcal{F}}}(\pi) > k$. Namely, when $\mathcal{B}_k(t) = 1$,

$$\Theta(\mathcal{B}_k)(t) = 0 \iff k < h_{\bar{\mathcal{F}}}^-(S, t). \tag{3.16}$$

Since $k < h_{\bar{\mathcal{F}}}^-(S, t)$ indicates $\mathcal{B}_k(t) = 1$, it follows that

$$\mathcal{B}_k(t) = 1 \wedge \Theta(\mathcal{B}_k)(t) = 0 \iff k < h_{\bar{\mathcal{F}}}^-(S, t). \tag{3.17}$$

Similarly, we have

$$\mathcal{B}_k(t) = 0 \wedge \Theta(\mathcal{B}_k)(t) = 0 \iff k \geq h_{\bar{\mathcal{F}}}^+(S, t). \tag{3.18}$$

It follows that

$$\Theta(\mathcal{B}_k)(t) = 0 \iff k < h_{\bar{\mathcal{F}}}^-(S, t) \vee k \geq h_{\bar{\mathcal{F}}}^+(S, t). \tag{3.19}$$

Note that $\Theta(\mathcal{B}_k)(t)$ is either 0 or 1. Therefore,

$$\Theta(\mathcal{B}_k)(t) = 1 \iff h_{\bar{\mathcal{F}}}^-(S, t) \leq k < h_{\bar{\mathcal{F}}}^+(S, t). \tag{3.20}$$

Then for each t,

$$\sum_{k=0}^{N-1} \Theta(\mathcal{B}_k)(t) = \sum_{k=0}^{N-1} \delta\left(h_{\bar{\mathcal{F}}}^-(S, t) \leq k < h_{\bar{\mathcal{F}}}^+(S, t)\right) \tag{3.21}$$

$$= h_{\bar{\mathcal{F}}}^+(S, t) - h_{\bar{\mathcal{F}}}^-(S, t) \tag{3.22}$$

$$= \Theta(\mathcal{F}). \tag{3.23}$$

□

On a Boolean map $\mathcal{B}_k^{\mathcal{F}}$, $\Theta_\varphi^S(\mathcal{B}_k^{\mathcal{F}})(t)$ is either 0 or 1. $\Theta_\varphi^S(\mathcal{B})(t) = 0$ iff $S \sim_k t$. Therefore, we have

Lemma 3.5 *Let $\mathcal{I} = (V, \mathcal{F})$ be a digital image, $\mathcal{F}$ be an N level feature map, and S be the seed set. Then*

$$\varphi_{\mathcal{F}}(S, t) = BMD(S, t). \tag{3.24}$$

In other words, the Boolean map distance is equivalent to $\varphi_{\mathcal{F}}$.

Different from the MBD $d_{\mathcal{F}}$, the distance function $\varphi_{\mathcal{F}}$ is not induced by a valid path length function, because π and π' can be two different paths in Eq. 3.6. As a result, the definition of $\varphi_{\mathcal{F}}$ is not as straightforward to interpret as the MBD. By showing the equivalence between $\varphi_{\mathcal{F}}$ and the Boolean map distance, the physical meaning of $\varphi_{\mathcal{F}}$ becomes more clear.

When the seed set is composed of all the image border pixels, the distance transform $\Theta_{\varphi}^{S}(\mathcal{B}_{k}^{\mathcal{F}})$ is equivalent to the activation map $\mathcal{M}(\mathcal{B}_{k}^{\mathcal{F}})$ in BMS. Thus, we have the following result.

Corollary 3.6 *Let $\mathcal{I} = (V, \mathcal{F})$ be an N-level feature map, and S be the set of image border pixels. In BMS, the sum of the activation maps $\{\mathcal{M}(\mathcal{B}_{k}^{\mathcal{F}})\}_{k=0}^{N-1}$ of $\mathcal{F}$ equals the distance transform $\Theta_{\varphi}^{S}(\mathcal{F})$:*

$$\sum_{k=0}^{N-1} \mathcal{M}(\mathcal{B}_{k}^{\mathcal{F}}) = \Theta_{\varphi}^{S}(\mathcal{F}). \tag{3.25}$$

This means that for a pixel t, $\varphi_{\mathcal{F}}(S, t)$ is proportional to the number of times that t is activated in the sampled Boolean maps from $\mathcal{F}$.

3.3.3   BMS Approximates the MBD Transform

In [40, 166], $\varphi_{\mathcal{F}}$ is introduced as an efficient lower bound approximation of the MBD $d_{\mathcal{F}}$. However, [40, 166] only provide an approximation error bound of $\varphi_{\mathcal{F}}$ when the seed set S is singleton. In what follows, we show an error bound result in the same form as in [40] for general connected seed sets.

Definition 3.7 Given a grayscale image $\mathcal{I} = (V, \mathcal{F})$, $\varepsilon_{\mathcal{I}} = \max_{i,j} |\mathcal{F}(i) - \mathcal{F}(j)|$ is the *maximum local difference*, where pixel i and j are arbitrary 8-adjacent neighbors.

Theorem 3.8 *Given a grayscale image $\mathcal{I} = (V, \mathcal{F})$ with 4-connected paths, if the seed set S is connected, then $\Theta_{\varphi}^{S}(\mathcal{F})$ approximates the MBD transform $\Theta_{d}^{S}(\mathcal{F})$ with errors no more than $2\varepsilon_{\mathcal{I}}$, i.e. for each pixel t,*

$$0 \leq d_{\mathcal{F}}(S, t) - \varphi_{\mathcal{F}}(S, t) \leq 2\varepsilon_{\mathcal{I}}. \tag{3.26}$$

The proof is provided in Appendix A.

Remark 3.9 The proved error bound will be quite loose if $\varepsilon_{\mathcal{F}}$ is large on a digital image. However, as discussed in [166], assuming the given imaging system smooths the scene by a point spread function, we can think of an "ideal" image as a continuous function defined on a continuous domain. Therefore, when the resolution goes to infinity, a digital image will approach the ideal image, and $\varepsilon_{\mathcal{F}}$ will go to zero. In this sense, the proved error bound guarantees the asymptotic accuracy of the approximation of the MBD.

Based on Lemma 3.5, we have the following result showing the relationship between our newly introduced BMD and the MBD.

Corollary 3.10 *Given a grayscale image* $\mathcal{I} = (V, \mathcal{F})$ *with 4-connected paths, if the seed set S is connected, then*

$$0 \leq d_{\mathcal{F}}(S, t) - BMD(S, t) \leq 2\varepsilon_{\mathcal{I}}. \tag{3.27}$$

Corollary 3.11 *Let $\mathcal{I} = (V, \mathcal{F})$ be an N-level feature map with 4-connected paths, and S be the set of image border pixels. In BMS, the sum of the activation maps of $\{\mathcal{B}_k^{\mathcal{F}}\}_{k=0}^{N-1}$ approximates the MBD transform $\Theta_d^S(\mathcal{F})$ with errors no more than $2\varepsilon_{\mathcal{I}}$.*

In other words, the surroundedness cue captured by BMS is closely related to the MBD to the image border. The MBD is shown to be more robust to blurring, noise and seed positioning for seeded image segmentation [40]. In Fig. 3.1, we use a synthetic test image to demonstrate the advantage of the MBD in capturing the surroundedness cue. In the test image, there are two surrounded square regions. The one with higher contrast is slightly blurred on its border. The image values range in [0, 1], and we have added Gaussian noise with $\sigma = 0.05$ to it. Given that the seeds are the image border pixels, the MBD and three other distance transforms are shown in Fig. 3.1. The path length functions for the three compared distance transforms are

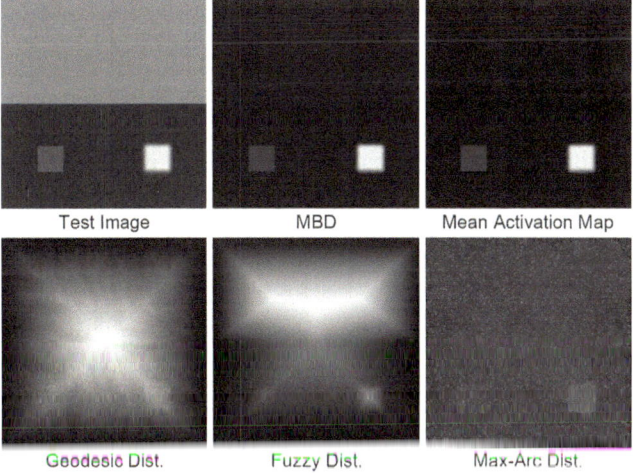

Fig. 3.1 In the test image (top left), there are two surrounded square regions. The one with higher contrast is slightly blurred on its border. The image values range in [0, 1], and we have added Gaussian noise with $\sigma = 0.05$ to it. Given that the seeds are the image border pixels, four types of distance transforms are shown: MBD (top middle), geodesic distance (bottom left), fuzzy distance (bottom middle), and max arc distance (bottom right). The mean activation map in BMS (top right) is computed using sample step $\delta = 0.02$. The values of all the maps are re-scaled for visualization. See text for more discussion

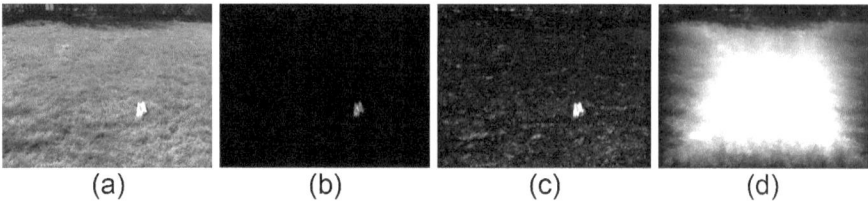

Fig. 3.2 (**a**) Test image. (**b**) Mean attention map in our formulation. (**c**) MBD transform. (**d**) Geodesic distance transform. See text for more discussion

listed as follows: (1) geodesic distance: $\tau_{\mathrm{GD}_{\mathcal{F}}}(\pi) = \sum_i |\mathcal{F}(\pi_i) - \mathcal{F}(\pi_{i-1})|$; (2) fuzzy distance [151]: $\tau_{\mathrm{FD}_{\mathcal{F}}} = \sum_i \frac{1}{2}(\mathcal{F}(\pi_i) + \mathcal{F}(\pi_{i-1}))$; (3) max-arc distance [57]: $\tau_{\mathrm{MA}_{\mathcal{F}}} = \max_i |\mathcal{F}(\pi_i) - \mathcal{F}(\pi_{i-1})|$.

As shown in Fig. 3.1, both the geodesic distance and the fuzzy distance lead to a brighter central area on the distance maps. This is because the effect of pixel value fluctuation can accumulate along a path for the geodesic distance and the fuzzy distance, due to the summational nature of these two distances. The max-arc distance is also sensitive to noise and blur and fails to capture the two square regions. In contrast, the MBD better handles the noise and the blur, and it properly measures the level of surroundedness of the two square regions. The mean activation map (or the sum of the activation maps) in BMS is very close to the MBD transform, as shown in Fig. 3.1.

As discussed above, the average of activation maps in BMS is equivalent to the BMD transform regarding the image border seed pixels. This straightforward computation of BMD can be used to approximate the MBD transform efficiently, which would otherwise require more complex computation [40]. Moreover, the Boolean map representation provides more flexibility as we can further process the Boolean maps for specific applications. For example, in our BMS formulation, the mean attention map can be thought of as a weighted average of the activation maps (after activation map splitting), where the weights are determined based on the rarity cue.

In Fig. 3.2, we show an example case where our Boolean map based formulation is advantageous over the MBD transform for saliency detection. Given a grayscale test image (Fig. 3.2a), the mean attention map (Fig. 3.2b) in our formulation successfully highlights the small white object in the scene. The MBD transform (Fig. 3.2c) detects the white object, but it is also quite responsive to the highly textured background. In addition, we also show the geodesic distance transform in Fig. 3.2d. Again, the geodesic distance is quite sensitive to the pixel value fluctuation in the background, leading to a bright central area. In [188], this phenomenon is identified as the *small weight accumulation problem* of the geodesic distance. Our mean attention map and the MBD transform do not suffer from this problem.

3.4 Distance Transform Algorithms

The relationship between the BMD and the MBD indicates that they can approximate each other when the seed set is connected. Therefore, any algorithms that can compute the distance transform for either of them can be used to approximate the other. In the following, we will summarize the major types of algorithms for this family of distance transforms.

Many image distance transforms can be solved by the *image foresting transform* [57], a generalization of Dijkstra's algorithm. Unfortunately, it is shown in [166] that the MBD transform cannot be exactly solved by Dijkstra-like algorithms due to the *non-smoothness* property [57] of the MBD. In [40], an exact algorithm for the MBD transform is proposed, but the computational cost of that algorithm is higher than that of Dijkstra's algorithm, asymptotically as well as in practice.

On the contrary, the exact solution to the BMD transform can be very efficient. As shown in previous sections, the discrete BMD is equivalent to $\varphi_{\mathcal{F}}$, which can be written as the difference between two functions:

$$\min_{\pi \in \Pi_{S,T}} \left(\max_{i \in \{0,1,...,k\}} \mathcal{F}(\pi(i)) \right), \tag{3.28}$$

and

$$\max_{\pi \in \Pi_{S,t}} \left(\min_{i \in \{0,1,...,k\}} \mathcal{F}(\pi(i)) \right). \tag{3.29}$$

Both of these functions are *path based distance functions*, and are *smooth* in the sense defined by Falcão et al. [57]. Therefore, distance transforms for each term can be computed in $\mathcal{O}(n \log n)$ operations using the *image foresting transform* [57], a generalization of Dijkstra's algorithm. In the case where the magnitude of the set of all intensities present in the image is bounded by a fixed integer, this can further be reduced to $\mathcal{O}(n)$ operations [57]. Therefore, the image foresting transform is an efficient way to compute the exact BMD transform and approximate the MBD transform.

Besides the image foresting transform algorithm, we can also use the Monte Carlo algorithm (see Algorithm 1) to compute the BMD transform. Although it is not as efficient as the image foresting transform, it has other advantages such as ease of implementation and runtime flexibility. The Monte Carlo algorithm can easily trade off accuracy for speed by reducing the sampling number.

Because of the relationship between the MBD and the BMD, the image foresting transform and the Monte Carlo algorithm can both be used for approximate MBD transform when the seed set is connected, providing very efficient and practical solutions to the MBD transform.

In the next chapter, we will introduce an iterative algorithm based on the raster scanning technique for approximate MBD transform. The raster scanning algorithm is cache friendly and is very fast in practice. We will show how this algorithm can be used for real-time salient region detection.

3.5   Conclusion

We have introduced the *Boolean map distance* (BMD), a pseudo-metric that measures the distance between elements in an image based on the probability that they belong to different components after thresholding the image by a randomly selected value. Formal definitions of the BMD have been given. The equivalence between the BMD and the φ mapping proposed by Strand et al. was proved and an error bound analysis between the BMD transform and the MBD transform was given, showing the relationship between the BMD and the MBD. We have also summarized available algorithms for computing distance transforms for the BMD and MBD.

Chapter 4
Efficient Distance Transform for Salient Region Detection

The goal of salient region detection is to compute a saliency map that highlights the regions of salient objects in a scene. Recently, this problem has received a lot of research interest owing to its usefulness in many computer vision applications, e.g. object detection, action recognition, and various image/video processing applications. Due to the emerging applications on mobile devices and large scale datasets, a desirable salient region detection method should not only output high quality saliency maps, but should also be highly computationally efficient. In this chapter, we address both the quality and speed requirements for salient region detection.

The *image boundary connectivity cue*, which assumes that background regions are usually connected to the image borders, is shown to be effective for salient region detection [187, 194, 201, 208]. To leverage this cue, previous methods, geodesic-distance-based [187, 208] or diffusion-based [84, 194], rely on a region abstraction step to extract superpixels. The superpixel representation helps remove irrelevant images details, and/or makes these models computationally feasible. However, this region abstraction step also becomes a speed bottleneck for this type of methods.

To boost the speed, we propose a method to exploit the image boundary connectivity cue without region abstraction. We use the *minimum barrier distance* (MBD) [40, 166] to measure a pixel's connectivity to the image boundary. Compared with the widely used geodesic distance, the MBD is much more robust to pixel value fluctuation. In contrast, the geodesic distance transform often produces a rather fuzzy central area when applied on raw pixels, due to the small-weight-accumulation problem observed in [187].

Since the exact algorithm for the MBD transform is not very efficient, we present FastMBD, a fast raster-scanning algorithm for the MBD transform, which provides a good approximation of the MBD transform in milliseconds, being two orders of magnitude faster than the exact algorithm [40]. Due to the *non-smoothness* property

© Springer Nature Switzerland AG 2019
J. Zhang et al., *Visual Saliency: From Pixel-Level to Object-Level Analysis*,
https://doi.org/10.1007/978-3-030-04831-0_4

Fig. 4.1 Sample saliency maps of several state-of-the-art methods (SO [208], AMC [84], HS [193], and SIA [35]) and methods with fast speed (HC [34], FT [1] and ours). Our method runs at about 80 FPS using a single thread, and produces saliency maps of high quality. Previous methods with similar speed, like HC and FT, usually cannot handle complex images well

[57] of MBD, error bound analysis of this kind of Dijkstra-like algorithm was previously regarded as difficult [40]. In this work, to the best of our knowledge, we present the first error bound analysis of a Dijkstra-like algorithm for the MBD transform.

The proposed salient region detection method runs at about 80 FPS using a single thread, and achieves comparable or better performance than the leading methods on four benchmark datasets. Compared with methods with similar speed, our method gives significantly better performance. Some sample saliency maps are shown in Fig. 4.1.

The main content of this chapter is summarized as follows.

1. We present FastMBD, a fast iterative MBD transform algorithm that is 100X faster than the exact algorithm, together with a theoretic error bound analysis.
2. We propose a fast salient region detection algorithm based on the MBD transform, which achieves state-of-the-art performance at a substantially reduced computational cost.

In addition, we provide an extension of the proposed method to leverage the appearance-based backgroundness cue [85, 108, 113]. This extension uses a simple and effective color space whitening technique, and it further improves the performance of our method, while still being at least one order of magnitude faster than all the other leading methods.

4.1   Fast Approximate MBD Transform

In this section, we present FastMBD, a fast raster scanning algorithm for the MBD transform, together with a new theoretic error bound result, which we believe should be useful beyond the application of salient region detection, e.g. in image/video segmentation and object proposal [99]. Note that we use notations slightly different from those in the previous chapter for the sake of consistency with our original paper.

4.1.1   Background: Distance Transform

The image distance transform aims at computing a distance map with respect to a set of background seed pixels. As a very powerful tool for geometric analysis of images, it has been a long-studied topic in computer vision [141].

Formally, we consider a 2-D single-channel digital image $\mathcal{I}$. A path $\pi = \langle \pi(0), \cdots, \pi(k) \rangle$ on image $\mathcal{I}$ is a sequence of pixels where consecutive pairs of pixels are adjacent. In this work, we consider 4-adjacent paths. Given a path cost function $\mathcal{F}$ and a seed set S, the distance transform problem entails computing a distance map $\mathcal{D}$, such that for each pixel t

$$\mathcal{D}(t) = \min_{\pi \in \Pi_{S,t}} \mathcal{F}(\pi), \tag{4.1}$$

where $\Pi_{S,t}$ is the set of all paths that connect a seed pixel in S and t.

The definition of the path cost function $\mathcal{F}$ is application dependent. In [187, 208], the geodesic distance is used for salient region detection. Given a single-channel image $\mathcal{I}$, the geodesic path cost function $\Sigma_{\mathcal{I}}$ is defined as

$$\Sigma_{\mathcal{I}}(\pi) = \sum_{i=1}^{k} |\mathcal{I}(\pi(i-1)) - \mathcal{I}(\pi(i))|. \tag{4.2}$$

where $\mathcal{I}(\cdot)$ denotes the pixel value. Recently, a new path cost function has been proposed in [166]:

$$\beta_{\mathcal{I}}(\pi) = \max_{i=0}^{k} \mathcal{I}(\pi(i)) - \min_{i=0}^{k} \mathcal{I}(\pi(i)). \tag{4.3}$$

The induced distance is called the *minimum barrier distance*, and it is shown to be more robust to noise and blur than the geodesic distance for seeded image segmentation [40, 166]. However, the exact algorithm for the MBD transform takes time complexity of $O(mn \log n)$ [40], where n is the number of pixels in the image and m is the number of distinct pixel values the image contains. In practice, an optimized implementation for the exact MBD transform can take about half a second for a 300 × 200 image [40].

4.1.2   Fast MBD Transform by Raster Scan

Inspired by the fast geodesic distance transform using the raster scanning technique [45, 176], we propose FastMBD, an approximate iterative algorithm for the MBD transform. In practice, FastMBD usually outputs a satisfactory result in a few iterations (see Sect. 4.1.3), and thus it can be regarded as having linear complexity in the number of image pixels. Like all raster scan algorithms, it is also cache friendly, so it is highly efficient in practice.

Algorithm 1 `FastMBD`

---

input   : image $\mathcal{I} = (I, V)$, seed set S, number of passes K
output   : MBD map $\mathcal{D}$
auxiliaries: $\mathcal{U}, \mathcal{L}$
 1: set $\mathcal{D}(x)$ to 0 for $\forall x \in S$; otherwise, set $\mathcal{D}(x)$ to ∞;
 2: set $\mathcal{L} \leftarrow \mathcal{I}$ and $\mathcal{U} \leftarrow \mathcal{I}$;
 3: **for** $i = 1 : K$ **do**
 4:    **if** mod $(i, 2) = 1$ **then**
 5:       RasterScan$(\mathcal{D}, \mathcal{U}, \mathcal{L}; \mathcal{I})$;
 6:    **else**
 7:       InvRasterScan$(\mathcal{D}, \mathcal{U}, \mathcal{L}; \mathcal{I})$;
 8:    **end if**
 9: **end for**

Algorithm 2 `RasterScan`$(\mathcal{D}, \mathcal{U}, \mathcal{L}; \mathcal{I})$

 1: **for** each x, which is visited in a raster scan order **do**
 2:    **for** each y in the masked area for x **do**
 3:       compute $\beta_{\mathcal{I}}(\mathcal{P}_y(x))$ according to Eq. 4.5;
 4:       **if** $\beta_{\mathcal{I}}(\mathcal{P}_y(x)) < \mathcal{D}(x)$ **then**
 5:          $\mathcal{D}(x) \leftarrow \beta_{\mathcal{I}}(\mathcal{P}_y(x))$;
 6:          $\mathcal{U}(x) \leftarrow \max\{\mathcal{U}(y), \mathcal{I}(x)\}$;
 7:          $\mathcal{L}(x) \leftarrow \min\{\mathcal{L}(y), \mathcal{I}(x)\}$;
 8:       **end if**
 9:    **end for**
10: **end for**

Similar to the raster scan algorithm for the geodesic or Euclidean distance transform, during a pass, we need to visit each pixel x in a raster scan or inverse raster scan order. Then each adjacent neighbor y in the corresponding half of neighborhood of x (see illustration in Fig. 4.2) will be used to iteratively minimize the path cost at x by

$$\mathcal{D}(x) \leftarrow \min \begin{cases} \mathcal{D}(x) \\ \beta_{\mathcal{I}}(\mathcal{P}(y) \cdot \langle y, x \rangle) \end{cases}, \tag{4.4}$$

where $\mathcal{P}(y)$ denotes the path currently assigned to the pixel y, $\langle y, x \rangle$ denotes the edge from y to x, and $\mathcal{P}(y) \cdot \langle y, x \rangle$ is a path for x that appends edge $\langle y, x \rangle$ to $\mathcal{P}(y)$.

Fig. 4.2 Illustration of the raster scan pass and the inverse raster scan pass. The green pixel is the currently visited pixel, and its masked neighbor area for 4-adjacency is shown in red

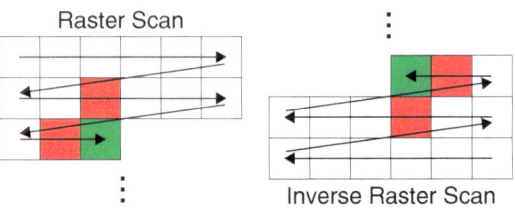

Let $\mathcal{P}_y(x)$ denote $\mathcal{P}(y) \cdot \langle y, x \rangle$. Note that

$$\beta_{\mathcal{T}}(\mathcal{P}_y(x)) = \max\{\mathcal{U}(y), \mathcal{T}(x)\} - \min\{\mathcal{L}(y), \mathcal{T}(x)\}, \qquad (1.5)$$

where $\mathcal{U}(y)$ and $\mathcal{L}(y)$ are the highest and the lowest pixel values on $\mathcal{P}(y)$, respectively. Therefore, the new MBD cost $\beta_{\mathcal{T}}(\mathcal{P}_y(x))$ can be computed efficiently by using two auxiliary maps $\mathcal{U}$ and $\mathcal{L}$ that keep track of the highest and the lowest values on the current path for each pixel.

Given the image $\mathcal{T}$ and the seed set S, the initialization of the distance map $\mathcal{D}$ and the auxiliary map $\mathcal{U}$ and $\mathcal{L}$ is described in Algorithm 1. Then the two subroutines, a raster scan pass and an inverse raster scan pass, are applied alternately to update $\mathcal{D}$ and the auxiliary maps, until the required number of passes is reached (see Algorithm 1). The subroutine for a raster scan is described in Algorithm 2. An inverse raster scan pass is basically the same as Algorithm 2, except that it enumerates the pixels in reverse order and uses different neighborhood masks, as illustrated in Fig. 4.2.

Each iteration of Algorithm 2 updates $\mathcal{U}$ and $\mathcal{L}$ accordingly when path assignment changes. Thus, at any state of Algorithm 1, $\mathcal{D}(x)$ is the MBD path cost of some path that connects the seed set S and x. It follows that $\mathcal{D}(x)$ is an upper bound of the exact MBD of x at any step. Algorithm 1 will converge, since each pixel value of $\mathcal{D}$ is non-negative and non-increasing during update. The converged solution will also be an upper bound of the exact MBD for each pixel.

4.1.3 Approximation Error Analysis

The update rule of FastMBD (Eq. 4.4) shares the same nature with Dijkstra's algorithm for solving the shortest path problem. However, it is shown in [166] that the MBD transform cannot be exactly solved by Dijkstra-like algorithms due to the *non-smoothness* property [57] of the MBD. Therefore, the converged solution of FastMBD generally does not equal the exact MBD transform. To facilitate discussion, we first introduce the following concept.

Definition 4.1 For an image $\mathcal{T}$, the maximum local difference $\varepsilon_{\mathcal{T}}$ is the maximum absolute pixel value difference between a pair of pixels that share an edge or a corner on $\mathcal{T}$.

For a lower-bound approximation algorithm to the MBD transform [166], it has been proved that the corresponding errors are bounded by $2\varepsilon_{\mathcal{T}}$ when the seed set is singleton [40, 166] or connected [202]. We remind the readers that $2\varepsilon_{\mathcal{T}}$ is a very loose bound, because for natural images, $\varepsilon_{\mathcal{T}}$ is usually above $127/255$. Nevertheless, such bounds can provide insight into the asymptotic behavior of an algorithm when an image approaches its continuous version, e.g. an *idealized image* in the continuous domain $\mathbb{R}^2$ [41], or a simple up-sampled version using bilinear interpolation.

The error-bound analysis techniques presented in the previous works [40, 166, 202] cannot be applied on a Dijkstra-like algorithm for the MBD transform. In what follows, we show a non-trivial sufficient condition when the converged solution of FastMBD is exact. We first introduce a slightly modified version of FastMBD, denoted as FastMBD*, which is the same as FastMBD except that the input image first undergoes a discretization step. In the discretization step, we use a rounding function $G(v; \varepsilon_\mathcal{I}) = \left\lfloor \frac{v}{\varepsilon_\mathcal{I}} \right\rfloor \varepsilon_\mathcal{I}$ to map each pixel value v to the largest integer multiples of $\varepsilon_\mathcal{I}$ below v. Then the discretized image $\widetilde{\mathcal{I}}$ is passed to Algorithm 1 to obtain a distance map for the original image $\mathcal{I}$.

Lemma 4.2 *Given an image $\mathcal{I}$ and a seed set S, let $d_{\beta_\mathcal{I}}(x)$ denote the MBD from S to the pixel x, and $\mathcal{D}$ denote the converged solution of FastMBD*. Assuming 4-adjacency, if the seed set S is connected,*[1] *then for each pixel x,*

$$|\mathcal{D}(x) - d_{\beta_\mathcal{I}}(x)| < \varepsilon_\mathcal{I}. \tag{4.6}$$

The proof of Lemma 4.2 is provided in Appendix B. The above error bound applies to a connected seed set, which is more general than the assumption of a single seed set in previous works [40, 166].

Corollary 4.3 *Let $\mathcal{I}$ be an image with integer pixel values. Assuming 4-adjacency, if the seed set is connected and $\varepsilon_\mathcal{I} = 1$, the converged solution of FastMBD is exact.*

Proof When $\varepsilon_\mathcal{I} = 1$, FastMBD will be the same as FastMBD* since $\widetilde{\mathcal{I}} = \mathcal{I}$. According to Lemma 4.2, $\mathcal{D}(x)$ will equal $d_{\beta_\mathcal{I}}(x)$ because $|\mathcal{D}(x) - d_{\beta_\mathcal{I}}(x)|$ must be an integer and it is less than 1.

The condition of $\varepsilon_\mathcal{I} = 1$ can be achieved by upsampling an integer-valued image by bilinear interpolation. Note that the MBD is quite robust to upsampling and blur [40], due to its formulation in Eq. 4.3. Thus, Corollary 4.3 can be regarded as a theoretic guarantee that FastMBD is exact in the limit.

Aside from the worst-case error bounds, in practice, mean errors and the convergence rates are of more importance. Therefore, we test FastMBD on the PASCAL-S dataset [110] and set all of the image boundary pixels as seeds. We convert the input images to grayscale, and compute the mean absolute approximation error of Algorithm 1 w.r.t. the exact MBD transform. The result is shown in Fig. 4.3. The average mean error drops below 10/255 after three scan passes (two forward and one backward passes), and each pass costs only about 2 ms for a 320×240 image. The proposed FastMBD using three passes is about 100X faster than the exact algorithm proposed in [40], and over 30X faster than the fastest approximation

---

[1]A set S is connected if any pair of seeds are connected by a path in S.

Fig. 4.3 Mean absolute
distance approximation error
against the number of
iterations K in the presented
fast algorithm for the MBD
transform. The pixel values of
the test images range between
0 and 255. The mean error
drops below 10/255 after
three scan passes

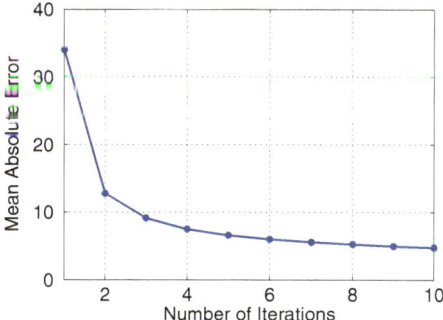

algorithm proposed in [40]. In the application of salient region detection, there
is no noticeable difference between FastMBD and the exact MBD transform in
performance.

4.2   Minimum Barrier Salient Region Detection

In this section, we describe an implementation of a system for salient region
detection that is based on FastMBD. Then an extension of our method is provided to
further leverage the appearance-based backgroundness cue. Lastly, several efficient
post-processing operations are introduced to finalize the salient map computation.

4.2.1   MBD Transform for Salient Region Detection

Similar to [187], to capture the image boundary connectivity cue, we set the pixels
along the image boundary as the seeds, and compute the MBD transform for each
color channel using FastMBD. Then the MBD maps for all color channels are pixel-
wise added together to form a combined MBD map $\mathcal{B}$, whose pixel values are further
scaled so that the maximum value is 1. We use three passes in FastMBD, as we find
empirically that increasing the number of passes does not improve performance.

An example is given in Fig. 4.4 to illustrate why the geodesic distance is less
favorable than the MBD in our case. We show the combined MBD map $\mathcal{B}$ (middle
top) and the combined geodesic distance map (middle bottom). The computation
of the geodesic distance map is the same as the MBD map, except that Eq. 4.2 is
used for the distance transform. Furthermore, as in [187], an adaptive edge weight
clipping method is applied on the geodesic distance map to alleviate the *small-
weight-accumulation* problem [187]. However, the geodesic distance map still has
a rather fuzzy central area, due to the fact that we compute the distance transform
on raw pixels instead of superpixels as in [187]. The MBD map does not suffer

Fig. 4.4 A test image is shown on the left. In the middle, the distance maps using the minimum barrier distance (top) and geodesic distance (bottom) are displayed. The corresponding resultant final saliency maps are shown in the last column. The geodesic distance map has a fuzzy central area due to its sensitivity to the pixel value fluctuation

from this problem. As a result, the final saliency map (right top) using the MBD suppresses the central background area more effectively than using the geodesic distance (right bottom).

4.2.2   Combination with Backgroundness Cue

We provide an extension of the proposed method by integrating the appearance-based backgroundness cue [85], which assumes that background regions are likely to possess similar appearance to the image boundary regions. This appearance-based cue is more robust when the salient regions touch the image boundary, and it is complementary to the geometric cue captured by the MBD map $\mathcal{B}$. Instead of using various regional appearance features as in previous works [85, 113], we present a more efficient way to leverage this backgroundness cue using color space whitening.

We compute an *image boundary contrast (IBC) Map* $\mathcal{U}$ to highlight regions with a high contrast in appearance against the image boundary regions. To do this, we consider four image boundary regions: (1) upper, (2) lower, (3) left, and (4) right. Each region is r pixels wide. For such a boundary region $k \in \{1, 2, 3, 4\}$, we calculate the mean color $\bar{\mathbf{x}}_k = [\bar{x}_1, \bar{x}_2, \bar{x}_3]$ and the color covariance matrix $\mathbf{Q}_k = [q_{ij}]_{3 \times 3}$ using the pixels inside this region. Then the corresponding intermediate IBC map $\mathbf{U}_k = [u_k^{ij}]_{W \times H}$ is computed based on the Mahalanobis distance from the mean color:

$$u_k^{ij} = \sqrt{\left(\mathbf{x}^{ij} - \bar{\mathbf{x}}_k\right) \mathbf{Q}_k^{-1} \left(\mathbf{x}^{ij} - \bar{\mathbf{x}}_k\right)^T}. \tag{4.7}$$

$\mathbf{U}_k$ is then normalized by $u_k^{ij} \leftarrow \frac{u_k^{ij}}{\max_{ij} u_k^{ij}}$, so that its pixel values lie in [0, 1]. The above formulation is equivalent to measuring the color difference in a whitened color space [142]. In a whitened color space, the Euclidean distance from the sample mean can better represent the distinctiveness of a pixel, because the coordinates of the whitened space are de-correlated and normalized.

Given the computed intermediate IBC maps $\{\mathbf{U}_k : k = 1, 2, 3, 4\}$ for the four image boundary regions, the final IBC map $\mathcal{U} = [u^{ij}]$ is computed by

$$u^{ij} = \left(\sum_{k=1}^{4} u_k^{ij} \right) - \max_k u_k^{ij}. \tag{4.8}$$

Compared with simply summing up all the intermediate IBC maps, the above formulation is more robust when one of the image boundary regions is mostly occupied by the foreground objects. Finally, we scale the values of $\mathcal{U}$ so that the maximum value is 1.

To integrate the IBC map $\mathcal{U}$ into our system, we pixel-wise add the MBD map $\mathcal{B}$ and the IBC map $\mathcal{U}$ together to form an enhanced map $\mathcal{B}^+ = \mathcal{B} + \mathcal{U}$. We find that although using $\mathcal{U}$ alone gives substantially worse performance than using $\mathcal{B}$, a simple linear combination of them consistently improves the overall performance.

4.2.3 Post-processing

We describe a series of efficient post-processing operations to enhance the quality of the final saliency map $\mathcal{S}$, given either $\mathcal{S} = \mathcal{B}$ or $\mathcal{S} = \mathcal{B}^+$. These operations do not add much computational burden, but can effectively enhance the performance for salient object segmentation.

Firstly, to smooth $\mathcal{S}$ while keeping the details of significant boundaries, we apply a morphological smoothing step on $\mathcal{S}$, which is composed of a reconstruction-by-dilation operation followed by a reconstruction-by-erosion [184]. The marker map for reconstruction by dilation (erosion) is obtained by eroding (dilating) the source image with a kernel of width δ. To make the smoothing level scale with the size of the salient regions, δ is adaptively determined by

$$\delta = \alpha \sqrt{s}, \tag{4.9}$$

where α is a predefined constant, and s is the mean pixel value on the map $\mathcal{B}$.

Secondly, similar to many previous methods [64, 158], to account for the center bias that is observed in many salient region detection datasets [15], we pixel-wise multiply $\mathcal{S}$ with a parameter-free centeredness map $\mathbf{C} = [c^{ij}]_{W \times H}$, which is defined as

$$c^{ij} = 1 - \frac{\sqrt{\left(i - \frac{H}{2}\right)^2 + \left(j - \frac{W}{2}\right)^2}}{\sqrt{\left(\frac{H}{2}\right)^2 + \left(\frac{W}{2}\right)^2}}. \tag{4.10}$$

Lastly, we scale the values of $\mathcal{S}$ so that its maximum value is 1, and we apply a contrast enhancement operation on $\mathcal{S}$, which increases the contrast between foreground and background regions using a sigmoid function:

$$f(x) = \frac{1}{1 + e^{-b(x-0.5)}}, \tag{4.11}$$

where b is a predefined parameter to control the level of contrast.

4.3 Experiments

Implementation In our implementation, input images are first resized so that the maximum dimension is 300 pixels. We set $\alpha = 50$ in Eq. 4.9, assuming the color values are in [0, 1]. We set $b = 10$ in Eq. 4.11. For our extended version, the width r of the border regions is set to 30. These parameters are fixed in the following experiments and we have found that, in practice, the performance of our algorithm is not sensitive to these parameter settings. An executable program of this implementation is available on our project website.[2]

Datasets To evaluate the proposed method, we use four large benchmark datasets: MSRA10K [1, 34, 112] (10,000 images), DUTOmron [194] (5168 images), ECSSD [193] (1000 images), and PASCAL-S [110] (850 images). Among these, the PASCAL-S and DUTOmron datasets are the most challenging, and the PASCAL-S dataset is designed to avoid the dataset design bias.

Compared Methods We denote our method and the extended version as MB and MB+, respectively. MB only uses the MBD map $\mathcal{B}$, and MB+ uses $\mathcal{B}^+$ which integrates the appearance-based backgroundness cue. We compare our method with several recently published methods: SO [208], AMC [84], SIA [35], HS [193], GS [187],[3] and RC [34]. We also include several methods with an emphasis on the speed performance: HC [33] and FT [1].

To demonstrate the advantages of MBD over the geodesic distance, we also evaluate a baseline method, denoted as GD. GD is the same as MB but uses the geodesic distance (Eq. 4.2) to compute the combined distance map $\mathcal{B}$ with the same post-processing applied. Adaptive edge weight clipping [187] is applied to alleviate the small-weight-accumulation problem. Parameters in the post-processing function are tuned to favor GD.

---

[2]http://www.cs.bu.edu/groups/ivc/fastMBD/.
[3]We use an implementation of GS provided by the authors of SO [208].

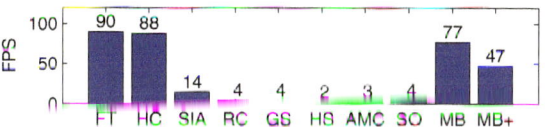

Fig. 4.5 Speed performance. Our methods MB and MB+ run at 77 and 47 FPS, respectively. While FT and HC are a bit faster, their accuracy is much lower (see Figs. 4.6 and 4.7)

4.3.1 Speed Performance

The speed performance of the compared methods is reported in Fig. 4.5. FT, HC, SIA, RC, and our methods are implemented in C, and the rest use C and Matlab. The evaluation is conducted on a machine with 3.2 GHz × 2 CPU and 12 GB RAM. We do not count I/O time, and do not allow processing multiple images in parallel. The test image size is the same as used in our methods (300 pixels in largest dimension) for all evaluated methods. Our method MB runs at about 80 FPS, which is comparable with the speed of FT and HC. Our extended version MB+ runs at 47 FPS, which is one order of magnitude faster than the state-of-the-art methods such as GS, HS, AMC, and SO.

4.3.2 Evaluation Using PR Curve

Similar to [1, 33, 84], we use Precision-Recall (PR) Curve to evaluate the overall performance of a method regarding its trade-off between the precision and recall rates. For a saliency map, we generate a set of binary images by thresholding at values in the range of [0, 1] with a sample step 0.05, and compute the precision and recall rates for each binary image. On a dataset, an average PR curve is computed by averaging the precision and recall rates for different images at each threshold value.

In the top row of Fig. 4.6, we show the PR curves for our methods MB and MB+, the baseline GD, and the methods with similar speed, FT and HC. MB outperforms GD, HC, and FT with a considerable margin across all datasets. The extended version MB+ further improves the performance of MB. On the most challenging dataset PASCAL-S, using the backgroundness cue only slightly increases the performance of MB+ over MB. Note that in many images in PASCAL-S the background is complex and the color contrast between the foreground and background is low.

In the bottom row of Fig. 4.6, we show the PR curves of our methods and the state-of-the-art methods. MB gives a better precision rate than SIA, RC, and GS over a wide range of recall rate across all datasets. Note that GS is based on the same image boundary connectivity cue, but it uses the geodesic distance transform on superpixels. The superior performance of MB compared with GS further validates the advantage of using the MBD over the geodesic distance. Compared with HS,

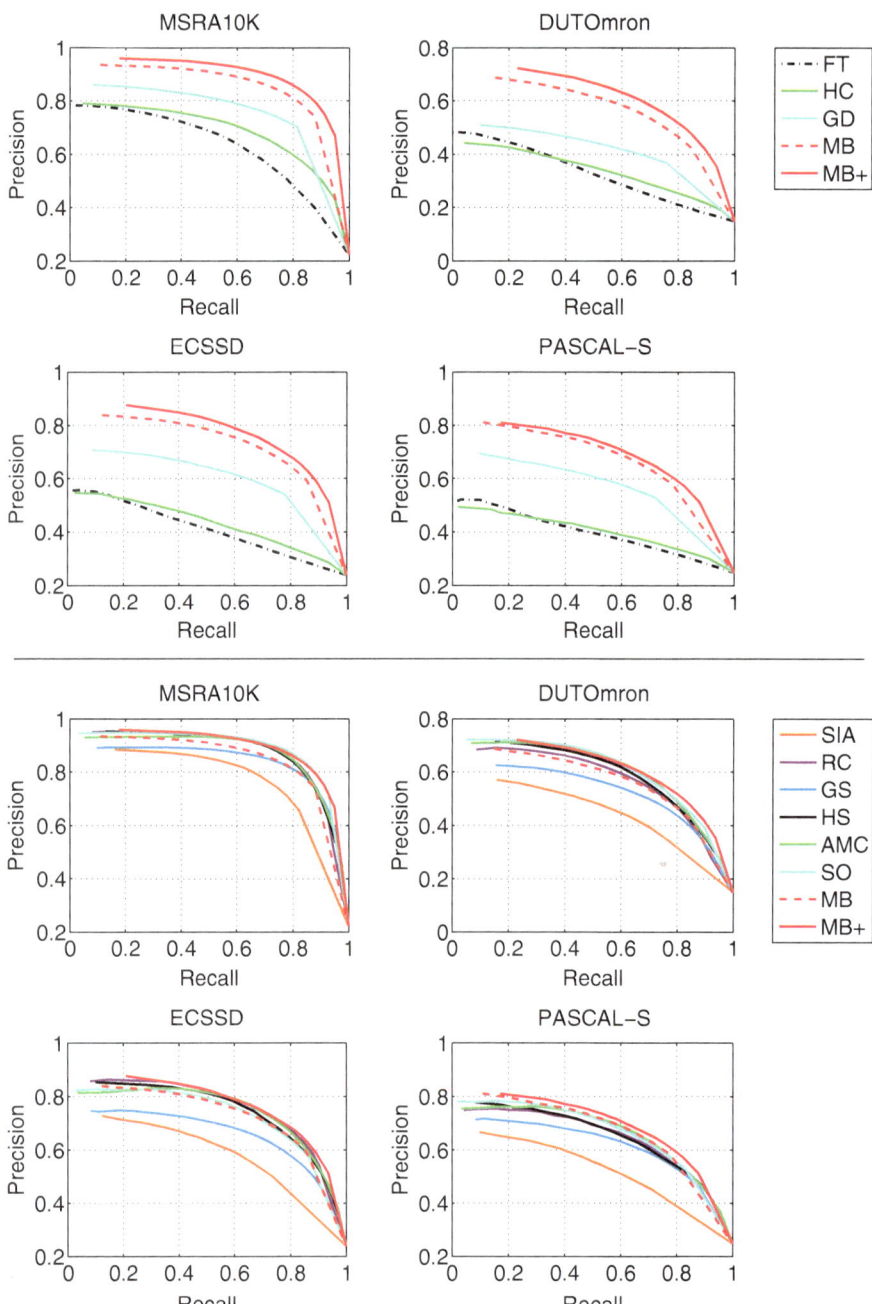

Fig. 4.6 Precision-recall curves of the compared methods. Our methods MB and MB+ significantly outperform methods that offer similar speed across all datasets (top row), and achieve state-of-the-art performance (bottom row). The PR curves of the baseline GD using the geodesic distance are significantly worse than its MBD counterpart MB, validating the advantage of the MBD over the geodesic distance in our application

AMC, and SO, MB achieves similar performance under the PR metric, while being over 25X faster. Our extended version MB+ consistently achieves state-of-the-art performance, and is over 10X faster than the other leading methods, such as HO, AMC, and SO.

4.3.3 Evaluation Using Weighted-F_β

To rank models, previous works use metrics such as area under the curve (AUC) [85, 113], average precision (AP) [113], and the F_β-measure [1, 33, 84]. However, as noted in [124], these metrics may not reliably evaluate the quality of a saliency map, due to the curve interpolation flaw, improper assumptions about the independence between pixels, and equal importance assignment to all errors. Therefore, we adopt the weighted-F_β metric proposed in [124], which suffers less from the aforementioned problems. We use the code and the default setting provided by the authors of [124]. For more information about the weighted-F_β metric, we refer the readers to [124].

The weighted-F_β scores are shown in Fig. 4.7. MB achieves significantly better scores than the methods with similar speed (FT and HC), and it compares favorably with SIA, RC, GS, HS, and AMC across all the datasets under the weighted-F_β metric. MB gives similar scores as the leading method SO on the MSRA10K and DUTOmron datasets, and it attains the best scores on the ECSSD and PASCAL datasets. Our extended version MB+ further improves the scores of MB, and attains the top weighted-F_β scores on all the datasets. The scores of the baseline method GD are substantially worse than those of MB, which is again consistent with our observation about the disadvantage of applying the geodesic distance on raw pixels.

To control the effect of post-processing on the ranking, we apply all possible combinations of the proposed post-processing steps for the other methods. In our post-processing function, there are three components (1) smoothing, (2) centeredness and (3) contrast enhancement. We try all possible $2^3 = 8$ combinations of these components (including none of the three and all of the three), and report the best score for each model. The results are shown in Fig. 4.8. We find that the proposed post processing routine substantially improves the scores of FT and HC, but it does not significantly improve or even degrade the scores of the other compared models. Controlling this factor does not lower the rankings of MB and MB+ on all the datasets.

Some sample saliency maps are shown in Fig. 4.9. Our methods MB and MB+ often give saliency maps with better visual quality than the other methods. The baseline GD tends to produce a rather fuzzy central area on the saliency map due to the small-weight-accumulation problem.

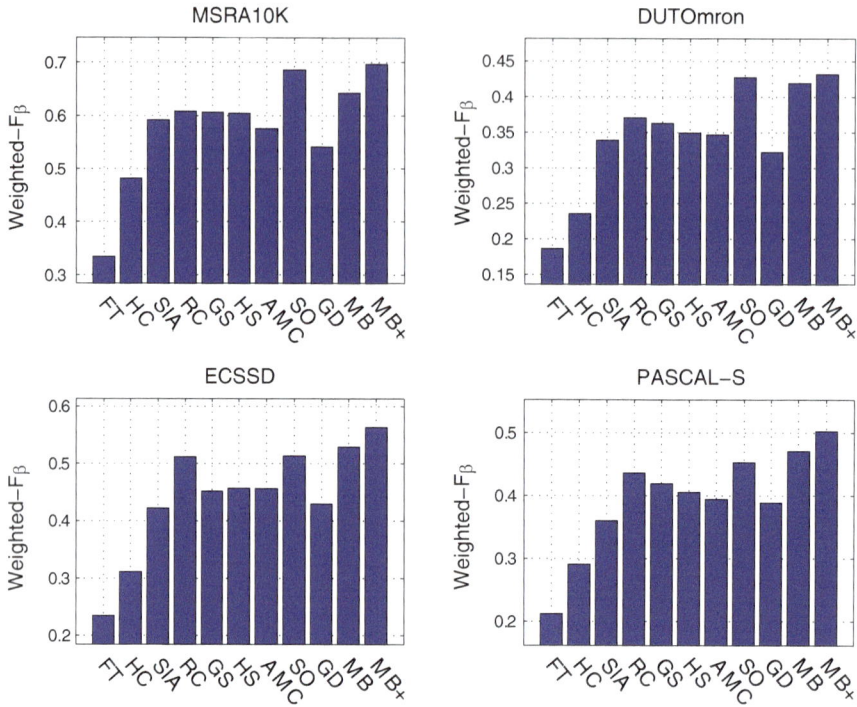

Fig. 4.7 Weighted-F_β scores of compared methods. Our methods MB and MB+ consistently attain comparable or better scores than the competitors

4.3.4 Limitations

A key limitation of the image boundary connectivity cue is that it cannot handle salient objects that touch the image boundary. In Fig. 4.10, we show two typical examples of this case. Our method MB fails to highlight the salient regions that are connected to the image boundary, because it basically only depends on the image boundary connectivity cue. Our extended version MB+, which further leverages the appearance-based backgroundness prior, can help alleviate this issue if the foreground region has a high color contrast against the image boundary regions (see the top right image in Fig. 4.10). However, when such backgroundness prior does not hold, e.g. in the second test image in Fig. 4.10, MB+ cannot fully highlight the salient region, either.

Fig. 4.8 Weighted-F_β scores when the effect of post-processing is controlled. In (**a**), we show the original scores for the compared methods. In (**b**), we show the scores after controlling the factor of post-processing (see text for more details). The post-processing improves the score of FT and HC, but does not significantly improve or even degrade the scores of the other compared models. As we can see, controlling this post-processing factor does not change the rankings of our methods

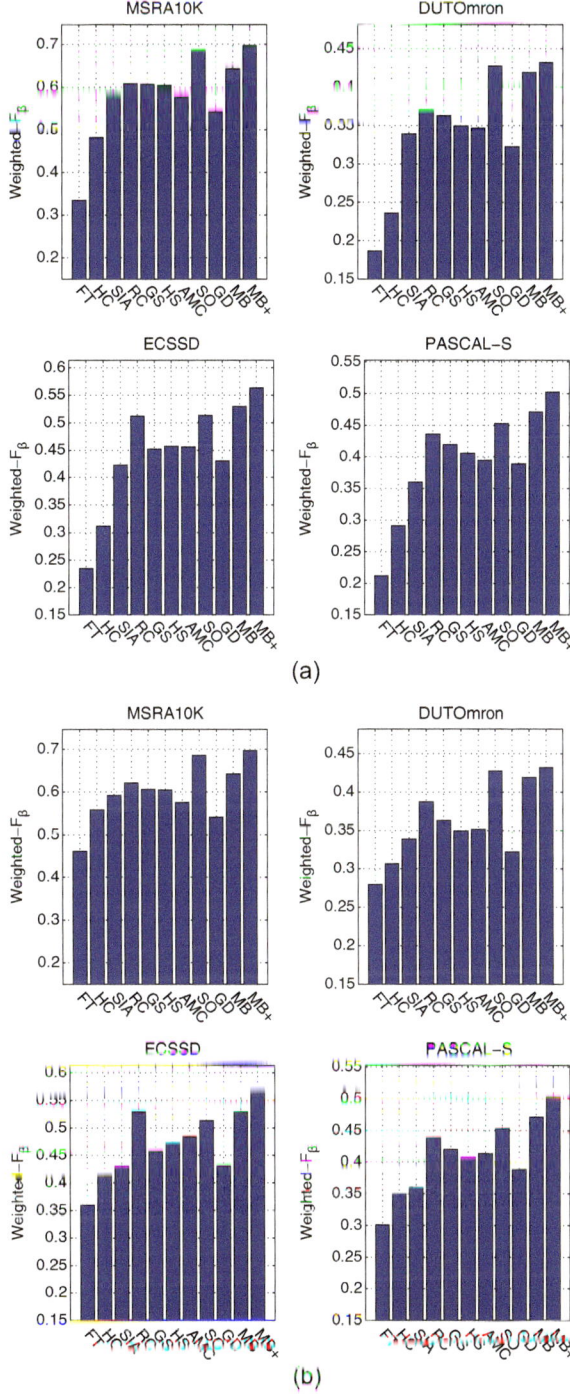

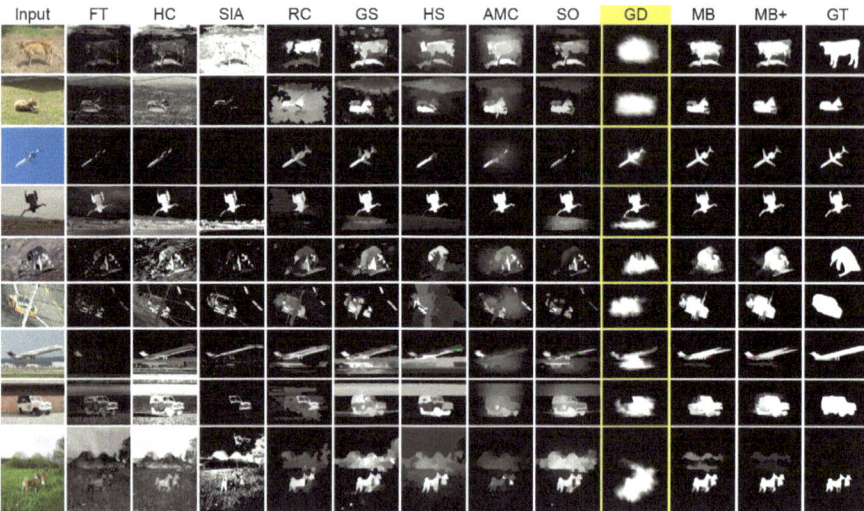

Fig. 4.9 Sample saliency maps of the compared methods. The baseline using the geodesic distance (GD) often produces a rather fuzzy central area, while our methods based on MBD (MB and MB+) do not suffer from this problem

Fig. 4.10 Some failure cases where the salient objects touch the image boundary

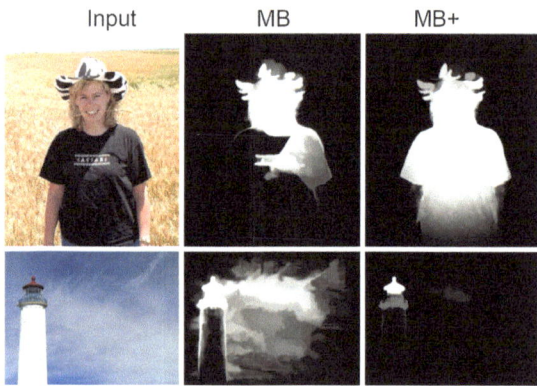

4.4 Conclusion

In this chapter, we presented FastMBD, a raster scanning algorithm to approximate the minimum barrier distance (MBD) transform, which achieves state-of-the-art accuracy while being about 100X faster than the exact algorithm. A theoretical error bound result was shown to provide insight into the good performance of such Dijkstra-like algorithms. Based on FastMBD, we proposed a fast salient region detection method that runs at about 80 FPS. An extended version of our method was also provided to further improve the performance. Evaluation was conducted on four benchmark datasets. Our method achieves state-of-the-art performance at a substantially smaller computational cost, and significantly outperforms the methods that offer similar speed

Our proposed minimum barrier salient region detection method is training-free, highly efficient, and easy to deploy. It is based on the image boundary connectivity assumption that background regions tend to be connected to the image boundary regions. While many previous methods leveraged this cue and achieved promising performance, our method does so in a different way. We introduce the minimum barrier distance (MBD), instead of the widely used geodesic distance, to measure the image boundary connectivity; and we propose a very fast approximate MBD transform algorithm. Thanks to the robustness of the MBD to pixel value fluctuation, our method can be directly applied on raw image pixels without additional super-pixel computations, which are often required by other methods. Due to its superior memory and computational efficiency, our method can be easily implemented on mobile platforms and embedded devices, e.g. digital cameras and eye trackers.

The image bounding connectivity assumption can be violated when foreground objects touch the image boundary. In this case, the MBD transform cannot well detect the salient region. To alleviate this issue, we propose to use the appearance based backgroundness cue, which is less sensitive to objects touching the image boundary. However, methods based on low-level cues are still very limited in their ability to handle unconstrained images, where foreground and background regions have complex appearance or there is no distinctive dominant object at all. Most existing salient region datasets lack such unconstrained images, so these challenges posed by unconstrained images are quite under-addressed. Li et al. [110] pointed out this negative aspect of dataset design bias and proposed a new dataset with many atypical images for salient region detection. It was shown that many state-of-the-art salient region detection methods perform poorly on this dataset. As demonstrated by the authors, to handle such unconstrained images, machine learning techniques that integrate multiple features can come to the rescue.

Part II
Object-Level Saliency

Chapter 5
Salient Object Subitizing

As early as the nineteenth century, it was observed that humans can effortlessly identify the number of items in the range of 1–4 by a glance [82] (see Fig. 5.1 for some examples). Since then, this phenomenon, later coined by Kaufman et al. as *subitizing* [90], has been studied and tested in various experimental settings [7, 121]. It is shown that identifying small numbers up to three or four is highly accurate, quick, and confident, while beyond this subitizing range, this sense is lost. Accumulating evidence also shows that infants and even certain species of animals can differentiate between small numbers of items within the subitizing range [46, 48, 67, 132]. This suggests that subitizing may be an inborn numeric capacity of humans and animals. It is speculated that subitizing is a preattentive and parallel process [48, 178, 185], and that it can help humans and animals make prompt decisions in basic tasks like navigation, searching, and choice making [66, 136].

Inspired by the subitizing phenomenon, we propose to study the problem of *salient object subitizing* (SOS), i.e. predicting the existence and the number (1, 2, 3, and 4+) of salient objects in an image without using any localization process. Solving the SOS problem can benefit many computer vision tasks and applications.

Knowing the existence and the number of salient objects without the expensive detection process can enable a machine vision system to select different processing pipelines at an early stage, making it more intelligent and reducing computational cost. For example, SOS can help a machine vision system suppress the object recognition process, until the existence of salient objects is detected, and it can also provide cues for generating a proper number of salient object detection windows for subsequent processing. Furthermore, differentiating between scenes with zero, a single and multiple salient objects can also facilitate applications like image retrieval, iconic image detection [11], image thumbnailing [38], robot vision [152], egocentric video summarization [104], snap point prediction [191], etc.

To study the SOS problem, we present a salient object subitizing image dataset of about 14K everyday images. The number of salient objects in each image was

© Springer Nature Switzerland AG 2019
J. Zhang et al., *Visual Saliency: From Pixel-Level to Object-Level Analysis*,
https://doi.org/10.1007/978-3-030-04831-0_5

Fig. 5.1 How fast can you tell the number of prominent objects in each of these images? It is easy for people to identify the number of items in the range of 1–4 by a simple glance. This "fast counting" ability is known as *subitizing*

Fig. 5.2 Sample images of the proposed SOS dataset. We collected about 14K everyday images, and use Amazon Mechanical Turk (AMT) to annotate the number of salient object of each image. The consolidated annotation is shown on the top of each image group. These images cover a wide range of content and object categories

annotated by Amazon Mechanical Turk (AMT) workers. The resulting annotations from the AMT workers were analyzed in a more controlled offline setting; this analysis showed a high inter-subject consistency in subitizing salient objects in the collected images. In Fig. 5.2, we show some sample images in the SOS dataset with the collected groundtruth labels.

We formulate the SOS problem as an image classification task, and aim to develop a method to quickly and accurately predict the existence and the number of generic salient objects in everyday images. We propose to use an end-to-end trained convolutional neural network (CNN) model for our task, and show that an implementation of our method achieves very promising performance. In particular, the CNN-based subitizing model can approach human performance in identifying images with no salient object and with a single salient object. We visualize the learned CNN features and show that these features are quite generic and discriminative for the class-agnostic task of subitizing. Moreover, we empirically validate the generalizability of the CNN subitizing model to unseen object categories.

To further improve the training of the CNN SOS model, we experiment with the usage of synthetic images. We generate a total of 20K synthetic images that contain different numbers of dominant objects using segmented objects and background images. We show that model pre-training using these synthetic images results in an absolute increase of more than 2% in average precision (AP) in identifying images with 2, 3, and 4+ salient objects, respectively. In particular, for images with three salient objects, our CNN model attains an absolute increase of about 6% in AP.

We demonstrate the application of our SOS method in salient object detection and image retrieval. For salient object detection, our SOS model can effectively suppress false object detections on background images and estimate a proper number of detections. By leveraging the SOS model, we attain an absolute increase of about 4% in F-measure over the state-of-the-art performance in unconstrained salient detection [204]. For image retrieval, we show that the SOS method can be used to handle queries with object number constraints.

In summary, our main contributions are:

1. We formulate the salient object subitizing (SOS) problem, which aims to predict the number of salient objects in an image without resorting to any object localization process.
2. We provide a large-scale image dataset for studying the SOS problem and benchmarking SOS models.
3. We present a CNN-based method for SOS, and propose to use synthetic images to improve the learned CNN model.
4. We demonstrate applications of the SOS method in salient object detection and image retrieval.

5.1 Related Work

For this new computer vision task, we review the following closely related areas.

Salient Object Detection Ideally, if a salient object detection method can well localize each salient object, then the number of objects can be simply inferred by counting the detection windows. However, many existing salient object detection methods assume the existence of salient objects, and they are mainly tested and optimized for images that contain a single dominant object, as observed in [15, 110]. Therefore, salient object detection methods often generate undesirable results on background images, and are prone to fail on images with multiple objects and complex background. Recently, [204] proposed a salient object detection method for unconstrained images. Although this method can handle complex images to some extent, we will show that the counting-by-detection approach is less effective than our subitizing method in predicting the number of salient objects.

Detecting the Existence of Salient Objects Only a few works address the problem of detecting the existence of salient objects in an image. [186] use a global feature based on several saliency maps to determine the existence of salient objects in thumbnail images. Their method assumes that an image either contains a single salient object or none. [152] use saliency histogram features to detect the existence of interesting objects for robot vision. It is worth noting that the testing images handled by the methods of [186] and [152] are substantially simplified compared to ours, and these methods cannot predict the number of salient objects.

Automated Object Counting There is a large body of literature about automated object counting based on density estimation [6, 106], object detection/segmentation [3, 129, 167], and regression [23, 24]. While automated object counting methods are often designed for crowded scenes with many objects to count, the SOS problem aims to discriminate between images with 0, 1, 2, 3, and 4+ dominant objects. Moreover, automated object counting usually focuses on a specific object category (e.g., people and cells), and assumes that the target objects have similar appearances and sizes in the testing scenario. On the contrary, the SOS problem addresses category-independent inference of the number of salient objects. The appearance and size of salient objects can vary dramatically from category to category, and from image to image, which poses a very different challenge than the traditional object counting problem.

Modeling Visual Numerosity Some researchers exploit deep neural network models to analyze the emergence of visual numerosity in human and animals [165, 210]. In these works, abstract binary patterns are used as training data, and the researchers study how the deep neural network model captures the number sense during either unsupervised or supervised learning. Our work looks at a more application-oriented problem, and targets at inferring the number of salient objects in natural images.

5.2 The SOS Dataset

We present the salient object subitizing (SOS) dataset, which contains about 14K everyday images. We first describe the collection of this dataset, and then provide a human labeling consistency analysis for the collected images. The dataset is available on our project website.[1]

5.2.1 Image Source

To collect a dataset of images with different numbers of salient objects, we gathered an initial set of images from four popular image datasets, COCO [111], ImageNet [148], VOC07 [56], and SUN [190]. Among these datasets, COCO, ImageNet, and VOC07 are designed for object detection, while SUN is for scene classification. Images from COCO and VOC07 often have complex backgrounds, but their content is limited to common objects and scenes. ImageNet contains a more diverse

---

[1] http://www.cs.bu.edu/groups/ivc/Subitizing/.

set of object categories, but most of its images have centered dominant objects with relatively simpler backgrounds. In the SUN dataset, many images are rather cluttered and do not contain any salient objects. We believe that combining images from different datasets can mitigate the potential data bias of each individual dataset.

This preliminary set is composed of about 30,000 images in total. There are about 5000 images from SUN, 5000 images from VOC07, 10,000 images are from COCO, and 10,000 images from ImageNet. For VOC07, the whole training and validation sets are included. We limited the number of images from the SUN dataset to 5000, because most images in this dataset do not contain obviously salient objects, and we do not want the images from this dataset to dominate the category for background images. The 5000 images were randomly sampled from SUN. For the COCO and ImageNet datasets,[2] we used the bounding box annotations to split the dataset into four categories for 1, 2, 3, and 4+, and then sampled an equal number of images from each category, in the hope that this can help balance the distribution of our final dataset.

5.2.2   Annotation Collection

We used the crowdsourcing platform Amazon Mechanical Turk (AMT) to collect annotations for our preliminary set of images. We asked the AMT workers to label each image as containing 0, 1, 2, 3, or 4+ prominent objects. Several example labeled images (shown in Fig. 5.3) were provided prior to each task as an instruction. We purposely did not give more specific instructions regarding some ambiguous cases for counting, e.g. counting a man riding a horse as one or two objects. We expected that ambiguous images would lead to divergent annotations.

Each task or HIT (Human Intelligence Task) was composed of five to ten images with a 2-min time limit, and the compensation was one to two cents per task. All the images in one task were displayed at the same time. The average completion time per image was about 4 s. We collected five annotations per image from distinct workers. About 800 workers contributed to this dataset. The overall cost for collecting the annotation is about 600 US dollars including the fees paid to the AMT platform.

A few images do not have a clear notion about what should be counted as an individual salient object, and labels on those images tend to be divergent. Cluttered images with multiple object categories are more likely to have inconsistent annotations, probably because the annotators may attend to different object categories. We show some of these images in Fig. 5.4. Cluttered images like these appear more frequently in VOC07 and COCO than in ImageNet. To remove these ambiguous images, we exclude images with fewer than four consensus labels, leaving about 14K images (out of 30K images) for our final SOS dataset. In Table 5.1, we show the joint distribution of images with respect to the labeled category and the original

---

[2]We use the subset of ImageNet images with bounding box annotations.

Fig. 5.3 Example labeled images for AMT workers. The number of salient objects is shown in the red rectangle on each image. There is a brief explanation below each image

Fig. 5.4 Sample images with divergent labels. These images are a bit ambiguous about what should be counted as an individual salient object. We exclude this type of images from the final SOS dataset

Table 5.1 Distribution of images in the SOS dataset

Category	COCO	VOC07	ImageNet	SUN	Total
0	616	311	371	1963	3261
1	2504	1691	1516	330	6041
2	585	434	935	76	2030
3	244	106	916	43	1309
4+	371	182	475	38	1066
Total	4320	2724	4213	2450	13,707

dataset. As expected, the majority of the images from the SUN dataset belong to the "0" category. The ImageNet dataset contains significantly more images with two and three salient objects than the other datasets.

5.2.3 Annotation Consistency Analysis

During the annotation collection process, we simplified the task for the AMT workers by giving them 2 min to label five images at a time. This simplification allowed us to gather a large number of annotations with reduced time and cost. However, the flexible viewing time allowed the AMT workers to look closely at these images, which may have had an influence over their attention and their answers to the number of salient objects. This leaves us with a couple important questions. Given a shorter viewing time, will labeling consistency among different subjects decrease? Moreover, will shortening the viewing time change the common answers to the number of salient objects? Answering these questions is critical in understanding our problem and dataset.

To answer these questions, we conducted a more controlled offline experiment based on common experimental settings in the subitizing literature [7, 121]. In this experiment, only one image was shown to a subject at a time, and this image was exposed to the subject for only 500 ms. After that, the subject was asked to tell the number of salient objects by choosing an answer from 0, 1, 2, 3, and 4+.

We randomly selected 200 images from each category according to the labels collected from AMT. Three subjects were recruited for this experiment, and each of them was asked to complete the labeling of all 1000 images. We divided that task into 40 sessions, each of which was composed of 25 images. The subjects received the same instructions as the AMT workers, except they were exposed to one image at a time for 500 ms. Again, we intentionally omitted specific instructions for ambiguous cases for counting.

Over 98% test images receive at least two out of three consensus labels in our experiment, and all three subjects agree on 84% of the test images. Table 5.2 shows the proportion of category labels from each subject that match the labels from AMT workers. All subjects agree with AMT workers on over 90% of sampled images. To see details of the labeling consistency, we show in Fig. 5.5 the averaged confusion matrix of the three subjects. Each row corresponds to a category label from the AMT workers, and in each cell, we show the average number (in the brackets) and percentage of images of category A (row number) classified as category B (column number). For categories 1, 2, and 3, the per-class accuracy scores are above 95%, showing that limiting the viewing time has little effect on the answers in these categories. For category 0, there is a 90% agreement between the labels from AMT workers and from the offline subitizing test, indicating that changing the viewing time may slightly affect the apprehension of salient objects. For category 4+, there is 78% agreement, and about 13% of images in this category are classified as category 0.

Table 5.2 Human subitizing accuracy in matching category labels from Mechanical Turk workers

	sbj. 1	sbj. 2	sbj. 3	Avg.
Accuracy	90%	92%	90%	91%

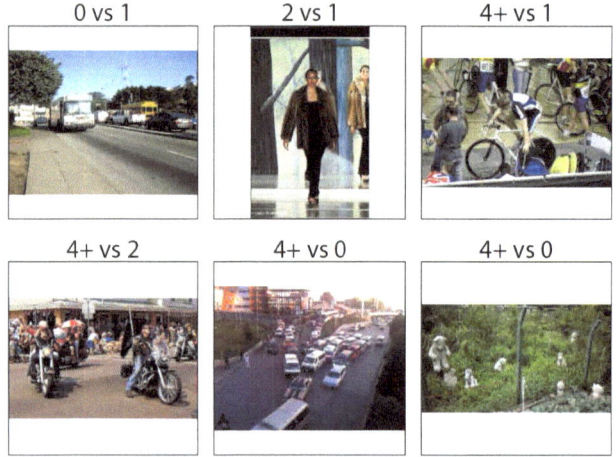

	0	1	2	3	4+
0	90% (179)	5% (9)	2% (3)	1% (2)	3% (6)
1	1% (2)	96% (191)	3% (5)	1% (1)	1% (1)
2	0	3% (6)	95% (189)	3% (5)	0
3	0	1% (1)	3% (5)	96% (191)	1% (2)
4+	13% (26)	3% (6)	4% (8)	2% (3)	78% (156)

Fig. 5.5 Averaged confusion matrix of our offline human subitizing test. Each row corresponds to a groundtruth category labeled by AMT workers. The percentage reported in each cell is the average proportion of images of the category A (row number) labeled as category B (column number). For over 90% images, the labels from the offline subitizing test are consistent with the labels from AMT workers

Fig. 5.6 Sample images that are consistently labeled by all three subjects in our offline subitizing test as a different category from what is labeled by the Mechanical Turk workers. Above each image, there is the AMT workers' label (left) vs the offline-subitizing label (right)

In Fig. 5.6, we show sample images that are consistently labeled by all three subjects in our offline subitizing test as a different category than labeled by AMT workers. We find some labeling discrepancy may be attributed to the fact that objects at the image center tend to be thought of as more salient than other ones given a short viewing time (see images in the top row of Fig. 5.6). In addition, some images with many foreground objects (far above the subitizing limit of 4) are labeled as 4+ by AMT workers, but they tend to be labeled as category 0 in our offline subitizing test (see the middle and right images at the bottom row in Fig. 5.6).

Despite the labeling discrepancy on a small proportion of the sampled images, limiting the viewing time to a fraction of a second does not significantly decrease the inter-subject consistency or change the answers to the number of salient objects on most test images. We thereby believe the proposed SOS dataset is valid. The per-class accuracy shown in Fig. 5.5 (percentage numbers in diagonal cells) can be interpreted as an estimate of the human performance baseline on our dataset.

5.3   Salient Object Subitizing by Convolutional Neural Network

Subitizing is believed to be a holistic sense of the number of objects in a visual scene. This visual sense can discriminate between the visual patterns possessed by different numbers of objects in an image [17, 42, 81, 121]. This inspires us to propose a learning-based discriminative approach to address the SOS problem, without resorting to any object localization or counting process. In other words, we aim to train image classifiers to predict the number of salient objects in an image.

Encouraged by the remarkable progress made by the CNN models in computer vision [62, 100, 139, 157], we use the CNN-based method for our problem. [62] suggest that given limited annotated data, fine-tuning a pre-trained CNN model can be an effective and highly practical approach for many problems. Thus, we adopt fine-tuning to train the CNN SOS model.

We use the GoogleNet architecture [172]. We fine-tune the GoogleNet model pre-trained on ImageNet [148] using Caffe [83]. The output layer of the pre-trained GoogleNet model is replaced by a fully connected layer which outputs a 5-D score vector for the five categories: 0, 1, 2, 3, and 4+. We use the Softmax loss and the SGD solver of Caffe to fine-tune all the parameters in the model. More training details are provided in Sect. 6.3.

5.3.1   Leveraging Synthetic Images for CNN Training

Collecting and annotating real image data is a rather expensive process. Moreover, the collected data may not have a balanced distribution over all the categories. In our SOS dataset, over 2/3 images belong to the "0" or "1" category. For categories with insufficient data, the CNN model training may suffer from overfitting and lead to degraded generalizability of the CNN model.

Leveraging synthetic data can be an economical way to alleviate the burden of image collection and annotation [80, 164, 171]. In particular, some recent works [80, 134] successfully exploit synthetic images to train modern CNN models for image recognition tasks. While previous works focus on generating realistic synthetic images (e.g., using 3D rendering techniques [134]) to train CNN models with zero or few real images data, our goal is to use synthetic images as an auxiliary source to improve the generalizability of the learned CNN model.

We adopt a convenient *cut-and-paste* approach to generate synthetic SOS image data. Given a number N in the range of 1–4, a synthetic image is generated by pasting N cutout objects on a background scene image. Cutout objects can be easily obtained from existing image datasets with segmentation annotations or image sources with isolated object photos (e.g., stock image databases). In this work, we use the public available salient object dataset THUS10000 [34] for generating cutout objects and the SUN dataset [190] as the source for background images. The THUS10000 dataset covers a wide range of object categories so that we can obtain sufficient variations in the shape and appearance of foreground objects.

In THUS10000, an image may contain multiple salient objects and some of them are covered by a single segmentation mask. To generate consistent synthetic SOS image data, we automatically filter out this type of images using the CNN SOS model trained on real data. To do this, we remove the images whose confidence scores for containing one salient object are less than 0.95. Similarly, we filter out the images with salient objects from the SUN dataset, using a score threshold of 0.95 for containing no salient object.

When generating a synthetic image, we randomly choose a background image and resize it to 256×256 regardless of its original aspect ratio. Then, we pick a cutout object and generate a reference object by resizing it to a randomly generated scale relative to 256 based on the largest dimension of the object. The reference scale is uniformly sampled in the range [0.4, 0.8]. After that, we apply random horizontal flipping and mild geometric transforms (scaling and rotation) on the reference object each time we paste a copy of it to a random position on the background image. Mild scalings are uniformly sampled in the range [0.85, 1.15] and mild rotations are uniformly sampled in the angular range $[-10, 10]$ degrees. The synthetic image contains N ($N \in [1, 4]$) copies of the same cutout object. Pasting different cutout objects together is empirically found inferior to our method, probably because some cutout objects may appear more salient than the other ones when they are put together, resulting in images that are visually inconsistent with the given number. Finally, we reject this image if any of the pasted objects is occluded by more than 50% of its area.

To generate synthetic images with multiple salient objects, we tried two approaches. The first one is randomly sampling different foreground objects for an image; the second one is using multiple copies of the same foreground object. We empirically find the former approach inferior to the latter one, probably because some cutout objects may appear more salient than the other ones when they are put together, resulting in images that are visually inconsistent with the given number. We also observe in the SOS dataset that if an image is annotated as having more than two salient objects, these objects tend to be in the same category (see sample images in Figs. 5.2 and 5.9). Note that we sampled our images from different sources

Fig. 5.7 Sample synthetic images with the given numbers of salient objects on the top. Although the synthetic images look rather unrealistic, they are quite visually consistent with the given numbers of salient objects. By pre-training the CNN SOS model on these synthetic images, we expect that the CNN model can better learn the intra-class variations in object category, background scene type, object position, and inter-object occlusion

(e.g., COCO and VOC07) to avoid dataset biases, but it turns out that images with multiple objects belonging to the same category received more consistent number annotations. We speculate that the foreground–background perception might get more clear if the objects in an image belong to the same category. Investigating the underlying cognitive mechanism can be an interesting future research problem.

Example synthetic images are shown in Fig. 5.7. Our synthetic images usually look rather unrealistic, since we do not consider any contextual constraints between scene types and object categories. For example, a motorbike can be pasted on a bedroom background; or birds' heads can appear inside a room. However, we expect that our CNN model should learn generic features for SOS irrespective of semantics of the visual scenes. Thus, these synthetic images may provide useful intra-class variations in object category, background scene type, as well as object position and inter-object occlusion.

Moreover, there can be boundary artifacts and color inconsistency between the foreground and the background. This kind of domain shift problem is very common in synthetic data generation. Applying dedicated blending and color harmonization techniques may reduce the artifacts in our synthetic data. However, to simplify our data generation pipeline, we do not apply any such techniques in this work. This allows us to examine how useful the synthetic data can be in the simplest form.

To properly leverage these synthetic data, we take the domain shift problem into account, and adopt a two-stage fine-tuning scheme. Namely, we fine-tune the CNN model on the synthetic data before fine-tuning on the real data. The two-stage fine-tuning scheme can be regarded as a domain adaptation process, which transfers the learned features from the synthetic data domain to the real data domain. Compared with combining the real and synthetic images into one training set, we find that our two-stage fine-tuning scheme works significantly better (see Sect. 5.4).

5.4 Experiments

5.4.1 Experimental Setting

For training and testing, we randomly split the SOS dataset into a training set of 10,966 images (80% of the SOS dataset) and a testing set of 2741 images.

CNN Model Training Details For fine-tuning the GoogleNet CNN model, images are resized to 256×256 regardless of their original aspect ratios. Standard data augmentation methods like horizontal flipping and cropping are used. We set the batch size to 32 and fine-tune the model for 8000 iterations. The fine-tuning starts with a learning rate of 0.001 and we multiply it by 0.1 every 2000 iterations. At test time, images are resized to 224×224 and the output softmax scores are used for evaluation.

For pre-training using the synthetic images, we generate 5000 synthetic images for each number in 1–4. Further increasing the number of synthetic images does not increase the performance. We also include the real background images (category "0") in the pre-training stage. The same model training setting is used as described above. When fine-tuning using the real data, we do not reset the parameters of the top fully connected layer, because we empirically find that it otherwise leads to slightly worse performance.

Compared Methods We evaluate our method and several baselines as follows.

- CNN_Syn_FT: The full model fine-tuned using the two-stage fine-tuning scheme with the real and synthetic image data.
- CNN_Syn_Aug: The model fine-tuned on the union of the synthetic and the real data. This baseline corresponds to the data augmentation scheme in contrast to the two-stage fine-tuning scheme for leveraging the synthetic image data. This baseline is to validate our two-stage fine-tuning scheme.
- CNN_FT: The CNN model fine-tuned on the real image data only.
- CNN_Syn: The CNN model fine-tuned on the synthetic images only. This baseline reflects how close the synthetic images are to the real data.
- CNN_wo_FT: The features of the pre-trained GoogleNet without fine-tuning. For this baseline, we fix the parameters of all the hidden layers during fine-tuning. In other words, only the output layer is fine-tuned.

Furthermore, we benchmark several commonly used image feature representations for baseline comparison. For each feature representation, we train a one-vs-all multi-class linear SVM classifier on the training set. The hyper-parameters of the SVM are determined via fivefold cross-validation.

- GIST. The GIST descriptor [177] is computed based on 32 Gabor-like filters with varying scales and orientations. We use the implementation by [177] to extract a 512-D GIST feature, which is a concatenation of averaged filter responses over a 4×4 grid.

- HOG. We use the implementation by [59] to compute HOG features. Images are first resized to 128 × 128, and HOG descriptors are computed on a 16 × 16 grid, with the cell size being 8 × 8. The HOG features of image cells are concatenated into a 7936-D feature. We have also tried combining HOG features computed on multi-scale versions of the input image, but this gives little improvement.
- SIFT with the Improved Fisher Vector Encoding (SIFT+IFV). We use the implementation by [26]. The codebook size is 256, and the dimensionality of SIFT descriptors is reduced to 80 by PCA. Hellinger's kernel and L2-normalization are applied for the encoding. Weak geometry information is captured by spatial binning using 1 × 1, 3 × 1, and 2 × 2 grids. To extract dense SIFT, we use the VLFeat [182] implementation. Images are resized to 256 × 256, and an 8 × 8 grid is used to compute a 8192-D dense SIFT feature, with a step size of 32 pixels and a bin size of 8 pixels. Similar to HOG, combining SIFT features of different scales does not improve the performance.
- Saliency map pyramid (SalPyr). We use a state-of-the-art CNN-based salient object detection model [206] to compute a saliency map for an image. Given a saliency map, we construct a spatial pyramid of an 8 × 8 layer and a 16 × 16 layer. Each grid cell represents the average saliency value within it. The cells of the spatial pyramid are then concatenated into a 320-D vector.

Evaluation Metric We use average precision (AP) as the evaluation metric. We use the implementation provided in the VOC07 challenge [56] to calculate AP. For each CNN-based method, we repeat the training for five times and report both the mean and the standard deviation (std) of the AP scores. This will give a sense of statistical significance when interpreting the difference between CNN baselines (Table 5.3).

Table 5.3 Average precision (%) of compared methods

	0	1	2	3	4+	Mean
Chance	27.5	46.5	18.6	11.7	9.7	22.8
SalPyr	46.1	65.4	32.6	15.0	10.7	34.0
HOG	68.5	62.2	34.0	22.8	19.7	41.4
GIST	67.4	65.0	32.3	17.5	24.7	41.4
SIFT+IFV	83.0	68.1	35.1	26.6	38.1	50.1
CNN_woFT	92.2 ± 0.2	84.4 ± 0.2	40.8 ± 1.9	34.1 ± 2.7	55.2 ± 0.6	61.3 ± 0.2
CNN_FT	**93.6 ± 0.3**	**93.8 ± 0.1**	75.2 ± 0.2	58.6 ± 0.8	71.6 ± 0.5	78.6 ± 0.2
CNN_Syn	79.2 ± 0.5	85.6 ± 0.2	37.4 ± 0.8	34.8 ± 2.6	33.0 ± 1.1	54.0 ± 0.6
CNN_Syn_Aug	92.1 ± 0.4	92.9 ± 0.1	75.0 ± 0.4	58.9 ± 0.6	69.8 ± 0.8	77.8 ± 0.3
CNN_Syn_FT	**93.5 ± 0.1**	**93.8 ± 0.2**	**77.4 ± 0.3**	**64.3 ± 0.2**	**73.0 ± 0.5**	**80.4 ± 0.2**

The best scores are shown in bold. The training and the testing are repeated for five times for all CNN-based methods, and mean and std of the AP scores are reported

5.4.2 Results

The AP scores of different features and CNN baselines are reported in Table 6.1. The baseline Chance in Table 6.1 refers to the performance of random guess. To evaluate the random guess baseline, we generate random confidence scores for each category, and report the average AP scores over 100 random trials.

All methods perform significantly better than random guess in all categories. Among manually crafted features, SalPyr gives the worst mean AP (mAP) score, while SIFT+IFV performs the best, outperforming SalPyr by 16 absolute percentage points in mAP. SIFT+IFV is especially more accurate than other non-CNN features in identifying images with 0 and 4+ salient objects.

The CNN feature without fine-tuning (CNN_wo_FT) outperforms SIFT+IFV by over ten absolute percentage points in mAP. Fine-tuning (CNN_FT) further improves the mAP score by 17 absolute percentage points, leading to a mAP score of 78.6%. CNN_wo_FT attains comparable performance to CNN_FT in identifying background images, while it is significantly worse than CNN_FT in the other categories. This suggests that the CNN feature trained on ImageNet is good for inferring the presence of salient objects, but not very effective at discriminating images with different numbers of salient objects.

Pre-fine-tuning using the synthetic images (CNN_Syn_FT) further boosts the performance of CNN_FT by about two absolute percentage points in mAP. The performance is improved in category "2", "3", and "4+", where training images are substantially fewer than categories "0" and "1". In particular, for category "3" the AP score is increased by about six absolute percentage points. The usefulness of the synthetic images may be attributed to the fact that they can provide more intra-class variations in object category, scene type, and the spatial relationship between objects. This is especially helpful when there is not enough real training data to cover the variations.

Using synthetic images alone (CNN_Syn) gives reasonable performance, a mAP score of 54.0%. It outperforms SIFT+IFV, the best non-CNN baseline trained on the real data. However, it is still much worse than the CNN model trained on the real data. This gives a sense of the domain shift between the real and the synthetic data. Directly augmenting the training data with the synthetic images does not improve and even slightly worsens the performance (compare CNN_Syn_Aug and CNN_FT in Table 6.1). We believe that this is due to the domain shift and our two-stage fine-tuning scheme can better deal with this issue.

Figure 5.8a shows the confusion matrix for our best method CNN_Syn_FT. The percentage reported in each cell represents the proportion of images of category A (row number) classified as category B (column number). The accuracy (recall) of category "0" and "1" is both about 93%, which is close to the human accuracy for these categories in our human subitizing test (see Fig. 5.5). For the remaining categories, there is still a considerable gap between human and machine performance. According to Fig. 5.8a, our SOS model tends to make mistakes by misclassifying an image into a nearby category. Sample results are displayed in

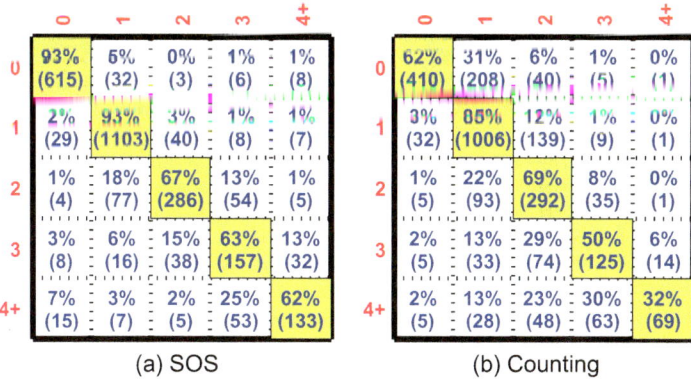

(a) SOS   (b) Counting

Fig. 5.8 Subitizing *vs.* counting. (**a**) Confusion matrix of our CNN SOS method CNN_Syn_FT. Each row corresponds to a groundtruth category. The percentage reported in each cell is the proportion of images of the category A (row number) labeled as category B (column number). (**b**) Confusion matrix of counting using the salient object detection method by [204]

Fig. 5.9 Sample results among the top 100 predictions for each category by our CNN SOS method CNN_Syn_FT. The images are listed in descending order of confidence. False alarms are shown with red borders and groundtruth labels at the top

Fig. 5.9. As we can see, despite the large variations in object appearance and image background, our SOS model gives promising performance. We also show the failure cases with top confidence scores in Fig. 5.10. We find that overlapping objects can cause difficulties for our model.

5.4.3 *Analysis*

To gain a better understanding of our SOS method, we further investigate the following questions.

0	1	2	3	4+

Fig. 5.10 Failure examples with highest confidence. Each column shows the top three failure cases for a category. Many failure cases are due to object occlusion or overlap. There are also a few ambiguous cases, e.g. the top one for "4+"

How Does Subitizing Compare to Counting? Counting is a straightforward way of getting the number of items. To compare our SOS method with a counting-by-detection baseline, we use a state-of-the-art salient object detection method designed for unconstrained images [204]. This unconstrained salient object detection method, denoted as USOD, leverages a CNN-based model for bounding proposal generation, followed by a subset optimization method to extract a highly reduced set of detection windows. A parameter of USOD is provided to control the operating point for the precision-recall tradeoff. We pick an operating point that gives the best F-score[3] on the multi-salient-object (MSO) dataset [200] in this experiment.

The confusion matrix of the counting baseline is shown in Fig. 5.8b. Compared with the SOS method (see Fig. 5.8a), the counting baseline performs significantly worse in all categories except "2". In particular, for "0" and "4+", the counting baseline is worse than the SOS method by about 30 absolute percentage points. This indicates that for the purpose of number prediction, the counting-by-detection approach can be a suboptimal option. We conclude that there are at least two reasons for this outcome. First, it is difficult to pick a fixed score threshold (or other equivalent parameters) of an object detection system that works best for all images. Even when an object detector gives a perfect ranking of window proposals for each image, the scores may not be well calibrated across different images. Second, the post-processing step for extracting detection results (e.g., non-maximum suppression)

---

[3]The F-score is computed as $\frac{2RP}{(R+P)}$, where R and P denote recall and precision respectively.

is based on the idea of suppressing severely overlapping windows. However, this spatial prior about detection windows can be problematic when significant inter-object occlusion occurs. In contrast, our SOS method bypasses the detection process and discriminates between different numbers of salient objects based on holistic cues.

How Does the CNN Model Architecture Affect the Performance? Besides GoogleNet, we evaluate another two popular architectures, AlexNet [100] and VGG16 [160]. The mAP scores with and without using synthetic images are summarized in Table 5.4 for each architecture. VGG16 and GoogleNet have very similar performance, while AlexNet performs significantly worse. Pre-training using synthetic images has a positive effect on all these architectures, indicating that it is generally beneficial to leverage synthetic images for this task. The baseline of AlexNet without synthetic image can be regarded as the best model reported by [200]. In this sense, our current best method using GoogleNet and synthetic image outperforms the previous best model by ten absolute percentage points. Note that the training and testing image sets used by [200] are subsets of the training and testing sets of our expanded SOS dataset. Therefore, the scores reported by [200] are not comparable to the scores in this paper.[4]

Does the Usage of Synthetic Images Reduce the Need for Real Data? To answer this question, we vary the amount of real data used in the training, and report the mAP scores in Table 5.5. We randomly sample 25% and 50% of the real data for training the model. This process is repeated for five times. When fewer real data

Table 5.4 Mean average precision (%) scores for different CNN architectures

	AlexNet	VGG16	GoogleNet
w/o Syn. Data	70.1 ± 0.2	77.5 ± 0.3	78.6 ± 0.2
With Syn. Data	71.6 ± 0.5	80.2 ± 0.3	80.4 ± 0.3

Training and test are run for five times and the mean and the std of mAP scores are reported

Table 5.5 The effect of using the synthetic images when different numbers of real data are used in CNN training

	w/o syn.	With syn.
25% real data	71.6 ± 0.2	76.3 ± 0.4
50% real data	73.3 ± 0.3	78.2 ± 0.4
100% real data	78.6 ± 0.2	80.4 ± 0.3

For each row, the same set of synthetic images are used. Training and test are run for five times and the mean and the std of mAP scores are reported. By using the synthetic images, competitive performance is attained even when the size of the real data is significantly reduced

---

[4]When evaluated on the test set used by [200], our best method GoogleNet_Syn_FT achieves a mAP score of 85.0%.

are used, the performance of our CNN SOS method declines much slower with the help of the synthetic images. For example, when only 25% real data are used, leveraging the synthetic images can provide an absolute performance gain of about 5% in mAP, leading to a mAP score of 76%. However, without using the synthetic images, doubling the size of the training data (50% real data) only achieves a mAP score of 75%. This suggests that we can achieve competitive performance at a much lower cost at data collection by leveraging the synthetic images.

What Is Learned by the CNN Model? By fine-tuning the pre-trained CNN model, we expect that the CNN model will learn discriminative and generalizable feature representations for subitizing. To visualize the new feature representations learned from our SOS data, we first look for features that are substantially distinct from the ones of the original network trained on ImageNet. For GoogleNet, we consider the output layer of the last inception unit (inception_5b/output), which has 1024 feature channels. For each feature channel of this layer, we use the maximum activation value on an image to rank the images in the SOS test set. We hypothesize that if two feature channels represent similar features, then they should result in similar image rankings. Given the i-th feature channel of this layer in GoogleNet_Syn_FT, we compute the maximum Spearman's rank correlation coefficient between its image ranking R_i and the image ranking $\widehat{R}_j$ using the j-th channel of the original GoogleNet:

$$S_i = \max_{j=1,2\cdots,1024} \rho(R_i, \widehat{R}_j), \tag{5.1}$$

where ρ denotes Spearman's rank correlation coefficient, whose range is $[-1, 1]$. A low value of S_i means that the i-th feature channel of our fine-tuned model gives a very different image ranking than any feature channels from the original CNN model. In our case, none of the values of S_i is negative. Figure 5.11a shows the histogram of S_i. We choose the feature channels with S_i values less than 0.3 as the most novel features learned from the SOS data.

After that, we visualize each of the novel feature channels by showing the top nine image patches in our SOS test set that correspond to the highest feature activations for that channel. The spatial resolution of inception_5b/output is 7×7. For an activation unit on the 7×7 map, we display the image patch corresponding to the receptive field of the unit. Since the theoretic receptive field of the unit is too large, we restrict the image patch to be 60% of the size ($0.6\,\mathrm{W} \times 0.6\,\mathrm{H}$) of the whole image.

Figure 5.11b shows the visualization results of some of the novel feature representations learned by our CNN SOS model. We find that these newly learned feature representations are not very sensitive to the categories of the objects, but they capture some general visual patterns related to the subitizing task. For example, in Fig. 5.11b, the feature corresponding to the first block is about a close-up face of either a person or an animal. Detecting a big face at this scale indicates that the image is likely to contain only a single dominant object. The feature corresponding to the second block is about a pair of objects appearing side by side, which is also

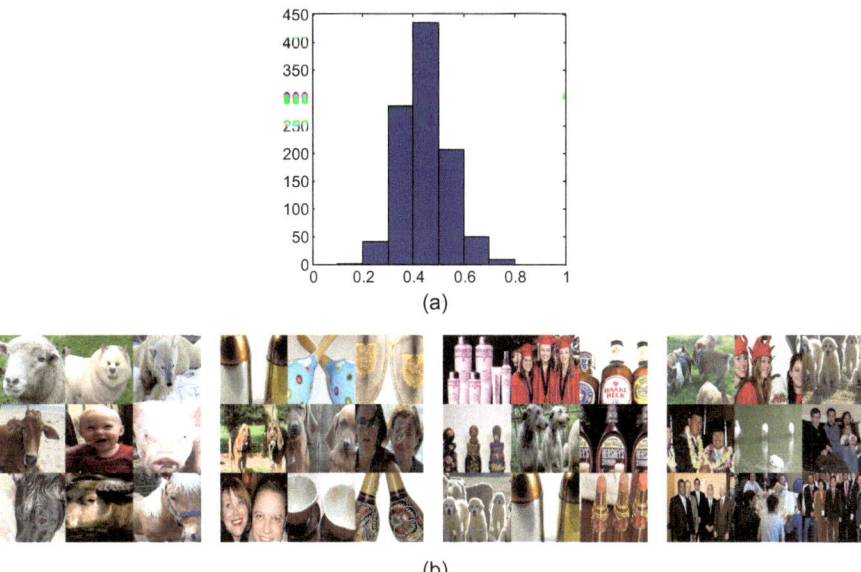

Fig. 5.11 Feature visualization of the inception_5b/output layer in our GoogleNet_Syn_FT model. We aim to visualize the new feature representations learned from our SOS data. (**a**) shows the histogram of S_i, which measures how distinct a feature channel of our model is from the feature representations of the original ImageNet model (see text for more details). Lower values of S_i indicate higher distinctness, and we choose those feature channels with $S_i < 0.3$ for visualization (**b**) shows the visualization of some new feature representations learned by our SOS model. Each block displays the top nine image patches in our SOS test set that correspond to the highest feature activations for a novel feature channel. These visualization results suggest that our CNN model has learned some category-independent and discriminative features for SOS. For example, the first block corresponds to a feature about a close-up face, and the second block shows a feature of a pair of objects appearing side by side

a discriminative visual pattern for identifying images with two dominant objects. These visualization results suggest that our CNN model has learned some category-independent and discriminative features for SOS.

In addition to the feature visualization, we also test the generalizability of our SOS model by manipulating the number of salient objects in an image using photoshop. We use photoshop's content-aware fill and move tools to add or remove salient objects in an image. We find that our SOS model responds well to these semantic image modifications. An example is shown in Fig. 5.12. This again shows that our SOS model learns meaningful features for subitizing.

How Does the SOS Method Generalize to Unseen Object Categories? We would like to further investigate how our CNN SOS model can generalize to unseen object categories. To get category information for the SOS dataset, we ask AMT

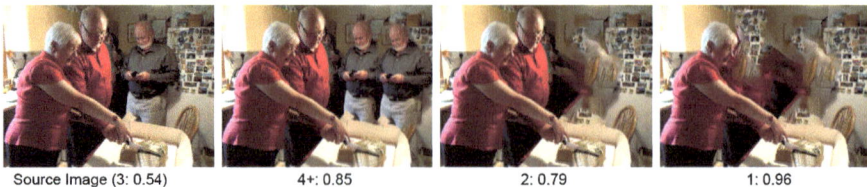

Source Image (3: 0.54)   4+: 0.85   2: 0.79   1: 0.96

Fig. 5.12 Effects of photo modification. We use photoshop's content-aware fill and content-aware move tools to remove and add salient objects in the image. The predicted category and the corresponding confidence score are displayed below each image. We find that our SOS model responds well to these semantic modifications. The above example is randomly sampled and is not cherry-picked

workers to label the categories of dominant objects for each image in our SOS dataset. We consider five categories: "animal," "food," "people," "vehicle," and "other." An image may contain multiple labels (e.g., an image with an animal and a person). For each image, we collect labels from three different workers and use the majority rule to decide the final labels.

To test the generalizability of our CNN model to unseen object categories, we use the Leave-One-Out (LOO) approach described as follows. Given category $\mathcal{A}$, we remove all the images with the label $\mathcal{A}$ from the original training set, and use them as the testing images. The original test images for "0" are also included. Two other baselines are provided. The first is a chance baseline, which refers to the performance of random guess. We generate random confidence scores for each category, and report the average AP scores over 100 random trials. Note that we have class imbalance in the test images, so the AP scores of random guess tend to be higher for categories with more images. The second baseline reflects the performance for category $\mathcal{A}$ when full supervision is available. We use fivefold cross-validation to evaluate this baseline. In each fold, $1/5$ of the images with the label $\mathcal{A}$ are used for testing, and all the remaining images are used for training. The average AP scores are reported. In this experiment, we do not use the synthetic images because they do not have category labels.

The results are reported in Table 5.6. For each category, the CNN model trained without that category (CNN-LOO) gives significantly better performance than the Chance baseline. This validates that the CNN model can learn category-independent features for SOS and it can generalize to unseen object categories to some extent. Training with full supervision (CNN-Full) further improves over CNN-LOO by a substantial margin, which indicates that it is still important to use a training set that covers a diverse set of object categories.

Table 5.6 Cross-category generalization test

		0	1	2	3	4+	Mean
Animal (4101)	Chance	16.6	33.6	21.1	12.6	8.8	22.5
	CNN-LOO	89.3 ± 0.2	87.2 ± 0.3	42.8 ± 1.0	36.9 ± 2.6	58.3 ± 1.0	62.9 ± 0.5
	CNN-Full	95.0 ± 1.7	94.8 ± 0.4	72.8 ± 2.0	57.9 ± 2.8	71.8 ± 4.0	78.5 ± 1.3
Food (372)	Chance	67.6	16.9	8.1	13.1	8.2	22.8
	CNN-LOO	95.7 ± 0.2	70.8 ± 1.3	50.3 ± 0.8	56.8 ± 1.3	39.7 ± 1.4	62.7 ± 0.5
	CNN-Full	97.7 ± 0.4	85.9 ± 7.2	61.1 ± 11.2	67.8 ± 12.4	62.8 ± 8.3	75.1 ± 4.1
People (3786)	Chance	17.5	50.7	21.7	10.9	13.1	22.8
	CNN-LOO	86.7 ± 0.3	84.9 ± 0.5	47.6 ± 0.5	31.6 ± 1.3	56.7 ± 1.2	61.5 ± 0.5
	CNN-Full	94.4 ± 1.3	94.8 ± 0.7	82.5 ± 1.0	62.8 ± 6.1	83.9 ± 2.8	83.7 ± 1.3
Vehicle (1150)	Chance	40.6	56.1	8.3	3.4	4.4	22.6
	CNN-LOO	91.0 ± 0.3	92.2 ± 0.3	42.4 ± 2.2	16.3 ± 0.9	47.4 ± 0.9	57.9 ± 0.4
	CNN-Full	96.1 ± 0.7	96.1 ± 0.7	62.2 ± 9.2	25.6 ± 14.2	55.4 ± 20.6	67.1 ± 6.4
Other (1401)	Chance	36.4	35.4	14.8	18.6	11.2	23.3
	CNN-LOO	87.0 ± 0.4	78.0 ± 0.7	56.7 ± 0.4	49.9 ± 0.9	50.2 ± 0.8	64.4 ± 0.4
	CNN-Full	93.4 ± 0.4	90.5 ± 2.5	70.8 ± 7.2	63.0 ± 3.2	60.2 ± 8.3	75.6 ± 2.8

The CNN-LOO refers to the AP scores (%) on the unseen object category. CNN-Full serves as an upper bound of the performance when the images of that object category are used in the training (see text for more details). The number following each category name is the number of images with that category label

5.5 Applications

In this section we demonstrate two applications of our CNN SOS model. In the following experiments, we use the best performing model CNN_wo_FT trained on the SOS dataset and our synthetic data.

5.5.1 Salient Object Detection

In this section, we demonstrate the usefulness of SOS for unconstrained salient object detection [204]. Unconstrained salient object detection aims to detect salient objects in unconstrained images where there can be multiple salient objects or no salient objects. Compared with the constrained setting, where there exists one and only one salient object, the unconstrained setting poses new challenges of handling background images and determining the number of salient objects. Therefore, SOS can be used to cue a salient object detection method to suppress the detection or output the right number of detection windows for unconstrained images. In the following, we demonstrate that SOS can improve unconstrained salient object detection performance by predicting an image-adaptive operating point.

Given a salient object detection method, we leverage our CNN SOS model by a straightforward approach. We assume that the salient object detection method provides a parameter (e.g., the threshold for the confidence score) for trade-off between precision and recall. We call this parameter as a PR parameter. For an image, we first predict the number of salient objects N using our CNN SOS model, then we use grid search to find such a value of the PR parameter that no more than N detection windows are output.

Dataset   Most existing salient object detection datasets lack background images or images containing multiple salient objects. In this experiment, we use the multi-salient-object (MSO) dataset [200]. The MSO dataset has 1224 images, all of which are from the test set of the SOS dataset, and it has a substantial proportion of images that contain no salient object or multiple salient objects.

Compared Methods   We test our SOS model on the unconstrained object detection method proposed (denoted as USOD) by [204], which achieves state-of-the-art performance on the MSO dataset. The baseline USOD method is composed of a CNN-based object proposal model and a subset optimization formulation for post-processing the bounding box proposals. We use an implementation provided by [204], which uses the GoogleNet architecture for proposal generation. The USOD method provides a PR parameter to control the number of detection windows. We use the predicted number by our SOS model to cue USOD, and denote this method as USOD+SOS. We also use the groundtruth number to show the upper-bound of the performance gain using subitizing, and denote this baseline as USOD+GT.

Evaluation Metrics   We report the precision, the recall, and the F-measure. The F-measure is calculated as $2\frac{PR}{P+R}$, where P and R denote the precision and the recall, respectively. For the baseline USOD method, we tune its PR parameter so that its F-measure is maximized.

Results   The results are reported in Table 5.7. Figure 5.13 shows the PR curve of USOD compared to the precision and recall rates of USOD+SOS and USOD+GT. As we can see, USOD+SOS significantly outperforms the baseline USOD, obtain-

Table 5.7 Salient object detection performance on the MSO dataset

		Prec.	Rec.	F-score
Full dataset	USOD	77.5	74.0	75.7
	USOD+SOS	79.6	79.5	79.5
	USOD+GT	83.9	81.7	82.8
Obj. Img.	USOD	78.0	81.0	79.4
	USOD+SOS	79.5	81.8	80.6
	USOD+GT	83.9	81.7	82.8

For the baseline USOD, we report its performance using the PR parameter that gives the optimal F-measure (%). We also report the performance of each method on a subset of the MSO dataset, which only contain images with salient objects (see Obj. Img.)

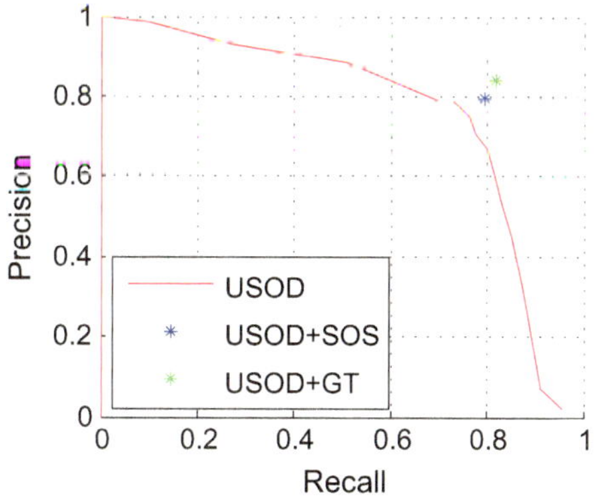

Fig. 5.13 Precision-Recall
curve of USOD, and the
performance of USOD+SOS
and USOD+GT

Table 5.8 Recognition
accuracy in predicting the
presence of salient objects on
the thumbnail image dataset
[186]

	[186]	Ours
Accuracy (%)	82.8	84.2

We show the fivefold cross
validation accuracy reported in
[186]. While our method is
trained on the MSO dataset, it
generalizes well to this other
dataset

ing an absolute increase of about 4% in F-measure. This validates the benefit of
adaptively tuning the PR parameter based on the SOS model. When the groundtruth
number of objects is used (USOD+GT), another absolute increase of 3% can be
attained, which is the upper bound for the performance improvement. Table 5.7
also reports the performance of each method on images with salient objects. On
this subset of images, using SOS improves the baseline USOD by about one
absolute percentage point in F-measure. This suggests that our CNN SOS model
is not only helpful for suppressing detections on background images, but is also
beneficial by determining the number of detection windows for images with salient
object.

Cross-Dataset Generalization Test for Identifying Background Images Detect
ing background images is also useful for tasks like salient region detection and
image thumbnailing [186]. To test how well the performance of our SOS model
generalizes to a different dataset for detecting the presence of salient objects in
images, we evaluate it on the web thumbnail image test set proposed by [106]. The
test set used by [186] is composed of 5000 thumbnail images from the Web, and
3000 images sampled from the MSRA-B [112] dataset. 50% of these images contain
a single salient object, and the rest contain no salient object. Images for MSRA-B
are resized to 130×130 to simulate thumbnail images [186].

In Table 5.8, we report the detection accuracy of our CNN SOS model, in
comparison with the fivefold cross-validation accuracy of the best model reported by
[186]. Note that our SOS model is trained on a different dataset, while the compared

model is trained on a subset of the tested dataset via cross validation. Our method outperforms the model of [186], and it can give fast prediction without resorting to any salient object detection methods. In contrast, the model of [186] requires computing several saliency maps, which takes over 4 s per image as reported by [186].

5.5.2   Image Retrieval

In this section, we show an application of SOS in content based image retrieval (CBIR). In CBIR, many search queries refer to object categories. It is useful in many scenarios that users can specify the number of object instances in the retrieved images. For example, a designer may search for stock images that contain two animals to illustrate an article about couple relationships, and a parent may want to search his/her photo library for photos of his/her baby by itself.

We design an experiment to demonstrate how our SOS model can be used to facilitate the image retrieval for number-object (e.g., "three animals") search queries. For this purpose, we implement a tag prediction system. Given an image, the system will output a set of tags with confidence scores. Once all images in a database are indexed using the predicted tags and scores, retrieval can be carried out by sorting the images according to the confidence scores of the query tags.

The Tag Prediction System   Our tag prediction system uses 6M training images from the Adobe Stock Image website.[5] Each training image has 30–50 user provided tags. We pick about 18K most frequent tags for our dictionary. In practice, we only keep the first five tags for an image as we empirically find that first few tags are usually more relevant. Noun Tags and their plurals are merged (e.g., "person" and "people" are treated as the same tag). We use a simple KNN-base voting scheme to predict image tags. Given a test image and a Euclidean feature space, we retrieve the 75 nearest neighbors in our training set using the distance encoded product quantization scheme of [72]. The proportion of the nearest neighbors that have a specific tag is output as the tag's confidence score.

The Euclidean feature space for the KNN system is learned by a CNN model. To learn the feature embedding, we use the 6M images in our retrieval database as the training data. For each image, we keep the first 20 tags and extract a *term frequency-inverse document frequency* (TF-IDF) representation out of these tags. Formally, we have

$$\mathbf{tfidf}(t, d, D) = \delta(t, d) \log(1 + \frac{N}{n_t}), \tag{5.2}$$

---

[5]https://stock.adobe.com.

where t is a word in the dictionary, d is the tag set associated with the given image, and D is the overall corpus containing all the tag sets associated with the images in our dataset. $\delta(t, d) = 1$ when $t \in d$ and 0 otherwise. N equals the number of images and n_t is the number of images that have the tag t.

Then the tag set associated with an image can be represented by a $\sim$18K-D vector where each entry is the TF-IDF value calculated by Eq. 5.2. We L2-normalize these tag TF-IDF vectors and group these vectors into 6000 clusters using k-means. We denote each cluster as a pseudo-class and assign the pseudo-class label to the images belonging to this cluster. To obtain the image feature embedding, we train a CNN image classifier for the pseudo-classes. We adopt the GoogleNet architecture [172] and train the model using the softmax loss function. Instead of training the model from scratch, we perform fine-tuning on the GoogleNet model pre-train on ImageNet. We set the batch size to 32 and fine-tune the model for 3 epochs. The fine-tuning starts with a learning rate of 0.01 and we multiply it by 0.1 after each epoch. After that, the output of the 1024D average pooling layer of the fine-tuned model is used as the feature embedding for the KNN image retrieval.

Dataset We use the publicly available NUS-WIDE dataset as our test set [39], which contains about 270K images. We index all the images of NUS-WIDE using our tag prediction system for all the tags of our dictionary. The NUS-WIDE dataset has the annotation of 81 concepts, among which we pick all the concepts that correspond to countable object categories as our base test queries (see Fig. 5.14 for the 37 chosen concepts). For a base test query, say "animal," we apply different test methods to retrieve images for four sub-queries, "one animal," "two animals," "three animals," and "many animals," respectively. Then all the retrieved images for "animal" by different test methods are mixed together for annotation. We ask three subjects to label each retrieved image as one of the four sub-queries or none of the sub-queries (namely, a five-way classification task). The subjects have no idea which test method retrieved which image. Finally, the annotations are consolidated by majority vote to produce the ground truth for evaluation.

Methods Given the tag confidence scores of each image by our tag prediction system, we use different methods to retrieve images for the number-object queries.

- Baseline. The baseline method ignores the number part of a query, and retrieves images using only the object tag.
- Text based method. This method treats each sub-query as the combination of two normal tags. Note that both the object tags and the number tags are included in our dictionary. We multiply the confidence scores of the object tag with the confidence scores of the number tag ("one", "two", "three" or "many"). Then the top images are retrieved according to the multiplied scores.
- SOS-based method. This method differs from the text-based method in that it replaces the number tag confidence score with the corresponding SOS confidence score. For a number tag "one/two/three/many," we use the SOS confidence score for 1/2/3/4+ salient object(s).

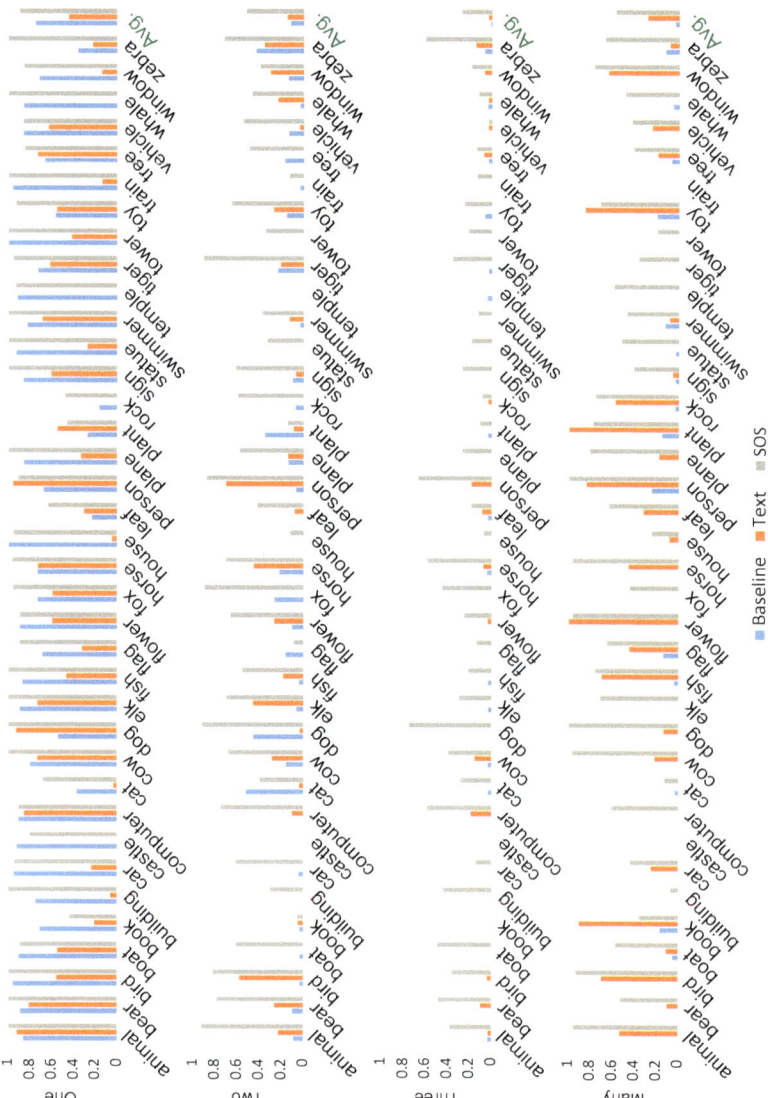

Fig. 5.14 nDCG scores for compared methods. For each object class, we use different methods to retrieve images of one/two/three/many object(s) of such class. The last column shows the average nDCG scores across different object classes

Evaluation Metric The widely used average precision (AP) requires annotation of the whole dataset for each number-object pair, which is too expensive. Therefore, we use the normalized discounted cumulative gain (nDCG) metric, which only looks at the top retrieved results. The nDCG is used in a recent image retrieval survey paper by [109] for benchmarking various image retrieval methods. The nDCG is formulated as

$$nDCG_h(t) = \frac{DCG_h(t)}{IDCG_h(t)}, \tag{5.3}$$

where t is the test query, $DCG_h(t) = \sum_{i=1}^{h} \frac{2^{rel_i}-1}{\log_2(i+1)}$, and rel_i denotes the tag relevance of the retrieved image at position i. In our case, rel_i is either 0 or 1. The $IDCG_h(t)$ is the maximum possible DCG up to position h. We retrieve 20 images for each method, so we set $h = 20$ and assume that there are at least 20 relevant images for each query.

Results The nDCG scores of our SOS-based method, the text-based method, and the baseline method are reported in Fig. 5.14. The SOS-based method gives consistently better average nDCG scores across queries for different numbers of objects, especially for the queries for more than one object. The scores of the SOS-based method for the group "three" are overall much lower than for the other groups. This is because the accuracy of our SOS is relatively lower for three objects. Moreover, there are many object categories that lack images with three objects, e.g. "statue," "rock," etc.

The baseline method gives pretty good nDCG scores for a single object, but for the other number groups, its performance is the worst. This reflects that images retrieved by a single object tag tend to contain only one dominant object. Note that it is often favorable that the retrieved images present a single dominant object of the searched category when no number is specified. When using SOS, the performance in retrieving images of one object is further improved, indicating it can be beneficial to apply SOS by default for object queries.

The text-based method is significantly worse than our SOS-based method across all number groups for most of the categories. Some exceptions are "many books" and "many plants." These categories tend to occur as textures of stuff when they are large in number, and thus might not be salient in the image. In these cases, using SOS may not be beneficial. The text-based method is significantly worse than our SOS-based method across all number groups. We observe that when a query has a number tag like "one," "two," and "three," the retrieved images by the text-based method tend to contain the given number of people. We believe that this is because these number tags often refer to the number of people in our training images. This kind of data bias obstructs the simple text-based approach to handling number-object queries. In contrast, our SOS-based method can successfully retrieve images for a variety of number-object queries, thanks to the category agnostic nature of our SOS formulation. Sample results of our SOS-based method are shown in Fig. 5.15.

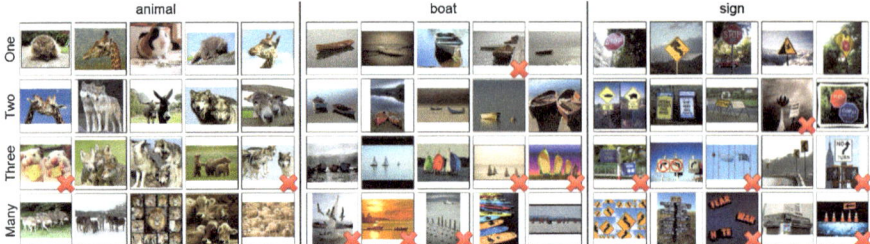

Fig. 5.15 Sample results of the SOS-based method for number-object image retrieval. The base object tags are shown above each block. Each row shows the top five images for a number group (one/two/three/many). Irrelevant images are marked by a red cross

5.5.3   Other Applications

The number of salient objects in a scene is correlated with the complexity of many scene analysis tasks. In [200] we showed that SOS can improve the efficiency for object proposal methods [137, 138] by adjusting the number of proposals according to the number of dominant objects. Beyond that, SOS has also been used as a scene complexity feature in engagement analysis for social media contents [159] and annotation agreement analysis in visual question answering [68]. Along this line, it will be interesting to further investigate the application of SOS in image captioning [192] and referring expressions generation [91], where the diversity and complexity of the textual description may have a significant correlation with the number of dominant objects in a scene.

5.6   Conclusion

In this chapter, we formulate the salient object subitizing (SOS) problem, which aims to predict the existence and the number of salient objects in an image using global image features, without resorting to any localization process. We collect an SOS image dataset, and present a convolutional neural network (CNN) model for this task. We leverage simple synthetic images to improve the CNN model training. Extensive experiments are conducted to show the effectiveness and generalizability of our CNN-based SOS method. We visualize that the features learned by our CNN model capture generic visual patterns that are useful for subitizing, and show how our model can generalize to unseen object categories. The usefulness of SOS is demonstrated in unconstrained salient object detection and content-based image retrieval. We show that our SOS model can improve the state-of-the-art salient object detection method, and it provides an effective solution to retrieving images by number-object queries.

The problem of salient object subitizing (SOS) is very different from previous visual saliency detection tasks, because it does not involve any localization process. SOS can be useful in visual systems that need quick, early prediction of the existence and the number of salient objects in an image without starting the computationally intensive detection or recognition process. This task is also very challenging, because it aims to do object counting on everyday images where objects may have dramatically different appearances, poses, and scales. According to our experiments, traditional image features are not successful for SOS, but our CNN-based SOS model provides very promising performance, thanks to the high expressiveness of the CNN features. In particular, our CNN subitizing model can process an image within milliseconds using a modern GPU, and the method can approach human performance in identifying images with no salient object or a single salient object.

In the training image set, there are fewer training images with multiple salient objects than images with a single salient object or no salient object. This data imbalance issue motivates us to use synthetic images to improve model training. However, our CNN SOS model still gives limited accuracy on images with multiple salient objects (see Fig. 5.8a). This is probably because predicting the number of salient objects when there exist multiple objects in an image is intrinsically more challenging than identifying images with a single salient object or no object.

Chapter 6
Unconstrained Salient Object Detection

In this chapter, we aim at detecting generic salient objects in unconstrained images, which may contain multiple salient objects or no salient object. Solving this problem entails generating a compact set of detection windows that matches the number and the locations of salient objects. To be more specific, a satisfying solution to this problem should answer the following questions:

1. (Existence) Is there any salient object in the image?
2. (Localization) Where is each salient object, if any?

These two questions are important not only in a theoretic aspect, but also in an applicative aspect. First of all, a compact and clean set of detection windows can significantly reduce the computational cost of the subsequent process (e.g., object recognition) applied on each detection window [63, 140]. Furthermore, individuating each salient object (or reporting that no salient object is present) can critically alleviate the ambiguity in the weakly supervised or unsupervised learning scenario [31, 89, 207], where object appearance models are to be learned with no instance level annotation.

However, many previous methods [1, 12, 34, 85, 113, 158, 203] only solve the task of foreground segmentation, i.e. generating a dense foreground mask (saliency map). These methods do not individuate each object. Moreover, they do not directly answer the question of existence. In this work, we will use the term *salient region detection* when referring to these methods, so as to distinguish from the *salient object detection* task solved by our approach, which includes individuating each of the salient objects, if there are any, in a given input image.

Some methods generate a ranked list of bounding box candidates for salient objects [60, 161, 195], but they lack an effective way to fully answer the questions of existence and localization. In practice, they just produce a fixed number of location proposals, without specifying the exact set of detection windows. Other salient object detection methods simplify the detection task by assuming the existence of

© Springer Nature Switzerland AG 2019
J. Zhang et al., *Visual Saliency: From Pixel-Level to Object-Level Analysis*,
https://doi.org/10.1007/978-3-030-04831-0_6

one and only one salient object [116, 170, 181]. This overly strong assumption limits their usage on unconstrained images.

In contrast to previous works, we present a salient object detection system that directly outputs a compact set of detections windows for an unconstrained image. Our system leverages the high expressiveness of a convolutional neural network (CNN) model to generate a set of scored salient object proposals for an image. Inspired by the attention-based mechanisms of [8, 102, 127], we propose an adaptive region sampling method to make our CNN model "look closer" at promising image regions, which substantially increases the detection rate. The obtained proposals are then filtered to produce a compact detection set.

A key difference between salient object detection and object class detection is that saliency greatly depends on the surrounding context. Therefore, the salient object proposal scores estimated on local image regions can be inconsistent with the ones estimated on the global scale. This intrinsic property of saliency detection makes our proposal filtering process challenging. We find that using the greedy non-maximum suppression (NMS) method often leads to sub-optimal performance in our task. To attack this problem, we propose a subset optimization formulation based on the *maximum a posteriori* (MAP) principle, which jointly optimizes the number and the locations of detection windows. The effectiveness of our optimization formulation is validated on various benchmark datasets, where our formulation attains about 12% relative improvement in average precision (AP) over the NMS approach.

In experiments, we demonstrate the superior performance of our system on three benchmark datasets: MSRA [112], DUT-O [194], and MSO [200]. In particular, the MSO dataset contains a large number of background/cluttered images that do not contain any dominant object. Our system can effectively handle such unconstrained images, and attains about 16–34% relative improvement in AP over previous methods on these datasets.

To summarize, the main contents are:

- A salient object detection system that outputs compact detection windows for *unconstrained* images,
- A novel MAP-based subset optimization formulation for filtering bounding box proposals,
- Significant improvement over the state-of-the-art methods on three challenging benchmark datasets.

6.1   Related Work

Few attempts have been made in detecting each individual salient object in unconstrained images. Therefore, we review several most relevant areas.

Salient Region Detection Salient region detection aims at generating a dense foreground mask (saliency map) that separates salient objects from the background of an image [1, 34, 85, 158, 193]. Some methods can detect regions of multiple salient objects [110, 119]. However, these methods do not separate each object.

Salient Object Localization   Given a saliency map, some methods find the best detection window based on heuristics [112, 116, 170, 181]. Various segmentation techniques are also used to generate binary foreground masks to facilitate object localization [65, 112, 115, 123]. A learning-based regression approach is proposed in [186] to predict a bounding box for an image. Most of these methods critically rely on the assumption that there is only one salient object in an image. In [112, 115], it is demonstrated that segmentation-based methods can localize multiple objects in some cases by tweaking certain parts in their formulation, but they lack a principled way to handle general scenarios.

Predicting the Existence of Salient Objects   Existing salient object/region detection methods tend to produce undesirable results on images that contain no dominant salient object [12, 186]. In [152, 186], a binary classifier is trained to detect the existence of salient objects before object localization. In [200], a salient object subitizing model is proposed to suppress the detections on background images that contain no salient object. While all these methods use a separately trained background image detector, we provide a unified solution to the problems of existence and localization through our subset optimization formulation.

Object Proposal Generation   Object proposal methods [2, 5, 25, 36, 180, 209] usually generate hundreds or thousands of proposal windows in order to yield a high recall rate. While they can lead to substantial speedups over sliding window approaches for object detection, these proposal methods are not optimized for localizing salient objects. Some methods [60, 161] generate a ranked list of proposals for salient objects in an image, and can yield accurate localization using only the top few proposals. However, these methods do not aim to produce a compact set of detection windows that exactly match the ground truth objects.

6.2   A Salient Object Detection Framework

Our salient object detection framework comprises two steps. It first generates a set of scored location proposals using a CNN model. It then produces a compact set of detections out of the location proposals using a subset optimization formulation. We first present the subset optimization formulation for bounding box filtering, as it is independent of the implementation of our proposal generation model, and can be useful beyond the scope of salient object detection.

For bounding box filtering, the greedy non-maximum suppression (NMS) is widely used due to its simplicity [2, 44, 59, 60]. Several limitations of greedy NMS are observed and addressed by [9, 49, 143, 147]. In [9], an improved NMS method is proposed for Hough transform based object detectors. Desai et al. [49] use a unified framework to model NMS and object class co-occurrence via context cueing. These methods are designed for a particular detection framework, which requires either part-based models or object category information. In [143], affinity propagation clustering is used for bounding box filtering. This method achieves more accurate bounding box localization, but slightly compromises average precision (AP). In [147], quadratic binary optimization is proposed to recover missing detections caused by greedy NMS. Unlike [143, 147], our subset optimization formulation aims to handle highly noisy proposal scores, where greedy NMS often leads to a poor detection precision rate.

Given a set of scored proposal windows, our formulation aims to extract a compact set of detection windows based on the following observations:

I. The scores of location proposals can be noisy, so it is often suboptimal to consider each proposal's score independently. Therefore, we jointly consider the scores and the spatial proximity of all proposal windows for more robust localization.

II. Severely overlapping windows often correspond to the same object. On the other hand, salient objects can also overlap each other to varying extents. We address these issues by *softly* penalizing overlaps between detection windows in our optimization formulation.

III. At the same time, we favor a compact set of detections that explains the observations, as salient objects are distinctive and rare in nature [51].

6.2.1 MAP-Based Proposal Subset Optimization

Given an image I, a set of location proposals $\mathbf{B} = \{b_i : i = 1 \ldots n\}$, and a proposal scoring function $\mathcal{S}$, we want to output a set of detection windows $\mathbf{O}$, which is a subset of $\mathbf{B}$. We assume each proposal b_i is a bounding box, with a score $s_i \triangleq \mathcal{S}(b_i, I)$. Given $\mathbf{B}$, the output set $\mathbf{O}$ can be represented as a binary indicator vector $(O_i)_{i=1}^{n}$, where $O_i = 1$ iff b_i is selected as an output.

The high-level idea of our formulation is to perform three tasks altogether: (1) group location proposals into clusters,)2) select an exemplar window from each cluster as an output detection, and (3) determine the number of clusters. To do so, we introduce an auxiliary variable $\mathbf{X} = (x_i)_{i=1}^{n}$. $\mathbf{X}$ represents the group membership for each proposal in $\mathbf{B}$, where $x_i = j$ if b_i belongs to a cluster represented by b_j. We also allow $x_i = 0$ if b_i does not belong to any cluster. Alternately, we can think that b_i belongs to the background. We would like to find the MAP solution w.r.t. the joint distribution $P(\mathbf{O}, \mathbf{X}|I; \mathbf{B}, \mathcal{S})$. In what follows, we omit the parameters $\mathbf{B}$ and $\mathcal{S}$ for brevity, as they are fixed for an image. According to Bayes' rule, the joint distribution under consideration can be decomposed as

$$P(\mathbf{O}, \mathbf{X}|I) = \frac{P(I|\mathbf{O}, \mathbf{X})P(\mathbf{O}, \mathbf{X})}{P(I)}. \tag{6.1}$$

For the likelihood term $P(I|\mathbf{O}, \mathbf{X})$, we assume that $\mathbf{O}$ is conditionally independent of I given $\mathbf{X}$. Thus,

$$P(I|\mathbf{O}, \mathbf{X}) = P(I|\mathbf{X})$$
$$= \frac{P(\mathbf{X}|I)P(I)}{P(\mathbf{X})}. \tag{6.2}$$

The conditional independence assumption is natural, as the detection set $\mathbf{O}$ can be directly induced by the group membership vector $\mathbf{X}$. In other words, representative windows indicated by $\mathbf{X}$ should be regarded as detection windows. This leads to the following constraint on $\mathbf{X}$ and $\mathbf{O}$:

Constraint 1 If $\exists x_i$ s.t. $x_i = j$, $j \neq 0$, then $b_j \in \mathbf{O}$.

To comply with this constraint, the prior term $P(\mathbf{O}, \mathbf{X})$ takes the following form:

$$P(\mathbf{O}, \mathbf{X}) = Z_1 P(\mathbf{X})L(\mathbf{O})\mathcal{C}(\mathbf{O}, \mathbf{X}), \tag{6.3}$$

where $\mathcal{C}(\mathbf{O}, \mathbf{X})$ is a constraint compliance indicator function, which takes 1 if Constraint 1 is met, and 0 otherwise. Z_1 is a normalization constant that makes $P(\mathbf{O}, \mathbf{X})$ a valid probability mass function. The term $L(\mathbf{O})$ encodes prior information about the detection windows. The definition of $P(\mathbf{O}, \mathbf{X})$ assumes the minimum dependency between $\mathbf{O}$ and $\mathbf{X}$ when Constraint 1 is met.

Substituting Eq. 6.2 and 6.3 into the RHS of Eq. 6.1, we have

$$P(\mathbf{O}, \mathbf{X}|I) \propto P(\mathbf{X}|I)L(\mathbf{O})\mathcal{C}(\mathbf{O}, \mathbf{X}). \tag{6.4}$$

Note that both $P(I)$ and $P(\mathbf{X})$ are cancelled out, and the constant Z_1 is omitted.

6.2.2   Formulation Details

We now provide details for each term in Eq. 6.4, and show the connections with the observations we made.

Assuming that the x_i are independent of each other given I, we compute $P(\mathbf{X}|I)$ as follows:

$$P(\mathbf{X}|I) = \prod_{i=1}^{n} P(x_i|I), \tag{6.5}$$

where

$$P(x_i = j|I) = \begin{cases} Z_2^i \lambda & \text{if } j = 0; \\ Z_2^i \mathcal{K}(b_i, b_j)s_i & \text{otherwise.} \end{cases} \tag{6.6}$$

Here Z_2^i is a normalization constant such that $\sum_{j=0}^n P(x_i = j|I) = 1$. $\mathcal{K}(b_i, b_j)$ is a function that measures the spatial proximity between b_i and b_j. We use window intersection over union (IOU) [56, 143] as $\mathcal{K}$. The parameter λ controls the probability that a proposal window belongs to the background. The formulation of $P(\mathbf{X}|I)$ favors representative windows that have strong overlap with many confident proposals. By jointly considering the scores and the spatial proximity of all proposal windows, our formulation is robust to individual noisy proposals. This addresses Observation **I**.

Prior information about detection windows is encoded in $L(\mathbf{O})$, which is formulated as

$$L(\mathbf{O}) = L_1(\mathbf{O})L_2(|\mathbf{O}|), \tag{6.7}$$

where

$$L_1(\mathbf{O}) = \prod_{i,j:i\neq j} \exp\left(-\frac{\gamma}{2}O_i O_j \mathcal{K}(b_i, b_j)\right). \tag{6.8}$$

$L_1(\mathbf{O})$ addresses Observation **II** by penalizing overlapping detection windows. Parameter γ controls the penalty level.

$L_2(|\mathbf{O}|)$ represents the prior belief about the number of salient objects. According to Observation **III**, we favor a small set of output windows that explains the observation. Therefore, $L_2(.)$ is defined as

$$L_2(N) = \exp(-\phi N), \tag{6.9}$$

where ϕ controls the strength of this prior belief.

Our MAP-based formulation answers the question of localization by jointly optimizing the number and the locations of the detection windows, and it also naturally addresses the question of existence, as the number of detections tends to be zero if no strong evidence of salient objects is found (Eq. 6.9). Note that $L(\mathbf{O})$ can also be straightforwardly modified to encode other priors regarding the number or the spatial constraints of detection windows.

6.2.3 Optimization

Taking the log of Eq. 6.4, we obtain our objective function:

$$f(\mathbf{O}, \mathbf{X}) = \sum_{i=1}^n w_i(x_i) - \phi|\mathbf{O}| - \frac{\gamma}{2} \sum_{i,j\in\tilde{\mathbf{O}}:i\neq j} \mathcal{K}_{ij}, \tag{6.10}$$

Algorithm 1 IncrementPass(**O**)

1: **V** ← **B** \ **O**;
2: **while V** ≠ ∅ **do**
3:     b^* ← arg max$_{b \in V}$ $h(\mathbf{O} \cup \{b\})$;
4:     **if** $h(\mathbf{O} \cup \{b^*\}) > h(\mathbf{O})$ **then**
5:         **O** ← **O** ∪ $\{b^*\}$;
6:         **V** ← **V** \ $\{b^*\}$;
7:     **else**
8:         **return**
9:     **end if**
10: **end while**

Algorithm 2 DecrementPass(**O**)

---

while O ≠ ∅ **do**
    b^* ← arg max$_{b \in \mathbf{O}}$ $h(\mathbf{O} \setminus \{b\})$
    if $h(\mathbf{O} \setminus \{b^*\}) > h(\mathbf{O})$ **then**
        └ **O** ← **O** \ $\{b^*\}$
    else
        └ **return**

---

where $w_i(x_i = j) \triangleq \log P(x_i = j|I)$ and $\mathcal{K}_{ij}$ is shorthand for $\mathcal{K}(b_i, b_j)$. $\tilde{\mathbf{O}}$ denotes the index set corresponding to the selected windows in **O**. We omit $\log \mathcal{C}(\mathbf{O}, \mathbf{X})$ in Eq. 6.10, as we now explicitly consider Constraint 1.

Since we are interested in finding the optimal detection set $\mathbf{O}^*$, we can first maximize over **X** and define our optimization problem as

$$\mathbf{O}^* = \arg\max_{\mathbf{O}} \left(\max_{\mathbf{X}} f(\mathbf{O}, \mathbf{X}) \right), \tag{6.11}$$

which is subject to Constraint 1. Given **O** is fixed, the subproblem of maximizing $f(\mathbf{O}, \mathbf{X})$ over **X** is straightforward:

$$\mathbf{X}^*(\mathbf{O}) = \arg\max_{\mathbf{X}} f(\mathbf{O}, \mathbf{X})$$

$$= \sum_{i-1}^{n} \max_{x_i \in \tilde{\mathbf{O}} \cup \{0\}} w_i(x_i). \tag{6.12}$$

Let $h(\mathbf{O}) \triangleq f(\mathbf{O}, \mathbf{X}^*(\mathbf{O}))$, then Eq. 6.11 is equal to an unconstrained maximization problem of the set function $h(\mathbf{O})$, as Constraint 1 is already encoded in $\mathbf{X}^*(\mathbf{O})$.

The set function $h(\mathbf{O})$ is submodular (see proof Appendix C) and the maximization problem is NP-hard [58]. We use a simple greedy algorithm to solve our problem. Our greedy algorithm starts from an empty solution set. It alternates between an incrementing pass (Algorithm 1) and a decrementing pass (Algorithm 2) until a local minimum is reached. The incrementing (decrementing) pass adds

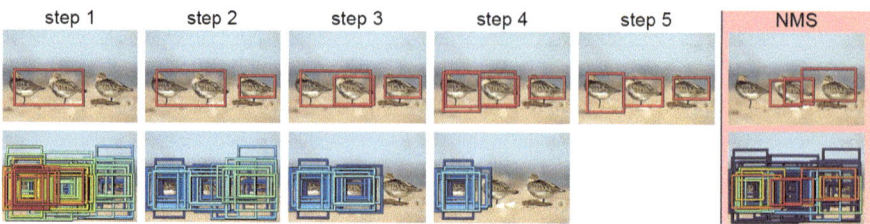

Fig. 6.1 In columns 1–5, we show step-by-step window selection results of our greedy algorithm. In the incrementing pass (steps 1–4), windows are selected based on their marginal gains *w.r.t.* Eq. 6.11. The window proposals of positive marginal gains are shown in the bottom row for each step. Warmer colors indicate higher marginal gains. The final step (step 5) removes the first selected window in the decrementing pass, because our formulation favors a small number of detection windows with small inter-window overlap. To contrast our method with greedy NMS, we show the top three output windows after greedy NMS using an IOU threshold of 0.4 (top). The scored proposals are shown in the bottom row of the figure

(removes) the element with maximal marginal gain to (from) the solution set until no more elements can be added (removed) to improve the objective function. Convergence is guaranteed, as $h(\mathbf{O})$ is upper-bounded and each step of our algorithm increases $h(\mathbf{O})$. An example of the optimization process is shown in Fig. 6.1.

In practice, we find that our greedy algorithm usually converges within two passes, and it provides reasonable solutions. Some theoretic approximation analyses for unconstrained submodular maximization [20, 58] may shed light on the good performance of our greedy algorithm.

6.2.4   Salient Object Proposal Generation by CNN

We present a CNN-based approach to generate scored window proposals $\{(b_i, s_i)\}_{i=1}^{n}$ for salient objects. Inspired by [55, 173], we train a CNN model to produce a fixed number of scored window proposals. As our CNN model takes the whole image as input, it is able to capture context information for localizing salient objects. Our CNN model predicts scores for a predefined set of exemplar windows. Furthermore, an adaptive region sampling method is proposed to significantly enhance the detection rate of our CNN proposal model.

Generating Exemplar Windows Given a training set with ground truth bounding boxes, we transform the coordinates of each bounding box to a normalized coordinate space, i.e. $(x, y) \rightarrow (\frac{x}{W}, \frac{y}{H})$, where W and H represents the width and height of the given image. Each bounding box is represented by a 4D vector composed of the normalized coordinates of its upper-left and bottom-right corners. Then we obtain K exemplar windows via K-means clustering in this 4D space. In our implementation, we set $K = 100$. We generate 100 exemplar windows by doing K-means clustering on the bounding box annotations of the SOS training set.

Adaptive Region Sampling The 100 exemplar windows only provide a coarse sampling of location proposals. To address this problem, the authors of [55, 173] propose to augment the proposal set by running the proposal generation method on uniformly sampled regions of an image. We find this uniformly sampling inefficient for salient object detection, and sometimes it even worsens the performance in our task (see Sect. 6.3).

Instead, we propose an adaptive region sampling method, which is in a sense related to the attention mechanism used in [8, 102, 127]. After proposal generation on the whole image, our model takes a closer glimpse at those important regions indicated by the global prediction. To do so, we choose the top M windows generated by our CNN model for the whole image, and extract the corresponding sub-images after expanding the size of each window by 2X. We then apply our CNN model on each of these sub-images to augment our proposal set. In our implementation, we set $M = 5$, and only retain the top ten proposals from each sub-image. This substantially speeds up the subsequent optimization process without sacrificing the performance.

The downside of this adaptive region sampling is that it may introduce more noise into the proposal set, because the context of the sub-images can be very different from the whole image. This makes the subsequent bounding box filtering task more challenging.

CNN Model Architecture and Training We use the VGG16 model architecture [160], and replace its fc8 layer with a 100-D linear layer followed by a sigmoid layer. Let $(c_i)_{i=1}^K$ denote the output of our CNN model. Logistic loss $\sum_i -y_i \log c_i - (1 - y_i) \log(1 - c_i)$ is used to train our model, where the binary label $y_i = 1$ iff the i-th exemplar window is the nearest to a ground truth bounding box in the 4D normalized coordinate space.

To train our CNN model, we use about 5500 images from the training split of the salient object subitizing (SOS) dataset [200]. The SOS dataset comprises unconstrained images with varying numbers of salient objects. In particular, the SOS dataset has over 1000 background/cluttered images that contain no salient objects, as judged by human annotators. By including background images in the training set, our model is expected to suppress the detections on this type of images. As the SOS dataset only has annotations about the number of salient objects in an image, we manually annotated object bounding boxes according to the number of salient objects given for each image. We excluded a few images that we found ambiguous to annotate.

We set aside 1/5 of the SOS training images for validation purpose. We first fine-tune the pre-trained VGG16 model on the ILSVRC-2014 object detection dataset [149] using the provided bounding box annotations, and then fine-tune it using the SOS training set. We find this two-stage fine-tuning gives lower validation errors than only fine-tuning on the SOS training set. The training images are resized to 224×224 regardless of their original dimensions. Training images are augmented by flipping and random cropping. Bounding box annotations that overlap with the cropping window by less than 50% are discarded. We use Caffe [83] to train the CNN model with a minibatch size of 8 and a fixed base learning rate of 10^{-4}. We fine-tune all the fully connected layers together with conv5_1, conv5_2, and conv5_3 layers by backpropagation. Other training settings are the same as in [160]. We fine-tune the model on the ILSVRC-2014 detection dataset for 230K iterations, when the validation error plateaus. Then we continue to fine-tune the model on the SOS training dataset for 2000 iterations, where the iteration number is chosen via fivefold cross validation. Fine-tuning takes about 20 h on the ILSVRC dataset, and 20 min on the SOS dataset using a single NVIDIA K40C GPU.

Our full system and the bounding box annotations of the SOS training set are available on our project website.[1]

6.3   Experiments

Evaluation Metrics Following [60, 161], we use the PASCAL evaluation protocol [56] to evaluate salient object detection performance. A detection window is judged as correct if it overlaps with a ground truth window by more than half of their union. We do not allow multiple detections for one object, which is different from the setting of [60]. Precision is computed as the percentage of correct predictions, and Recall is the percentage of detected ground truth objects. We evaluate each method by (1) Precision-Recall (PR) curves, which are generated by varying a parameter for each method (see below), and (2) Average Precision (AP), which is computed by averaging precisions on an interpolated PR curve at regular intervals (see [56] for details).

Precision-Recall Tradeoff As our formulation does not generate scores for the detection windows, we cannot control the PR tradeoff by varying a score threshold. Here we provide a straightforward way to choose an operating point of our system. By varying the three parameters in our formulation, λ, γ, and ϕ, we find that our system is not very sensitive to ϕ in Eq. 6.9, but responds actively to changes in λ and γ. λ controls the probability of a proposal window belonging to the background (Eq. 6.6), and γ controls the penalty for overlapping windows (Eq. 6.8). Thus, lowering either λ or γ increases the recall. We couple λ and γ by setting $\gamma = \alpha\lambda$, and fix ϕ and α in our system. In this way, the PR curve can be generated by varying λ. The parameters ϕ and α are optimized by grid search on the SOS training split. We fix ϕ at 1.2 and α at 10 for all experiments.

Compared Methods Traditional salient region detection methods [1, 34, 85, 158, 193] cannot be fairly evaluated in our task, as they only generate saliency maps without individuating each object. Therefore, we mainly compare our method with

---

[1]http://www.cs.bu.edu/groups/ivc/SOD/.

two state of the-art methods, SC [60] and LBI [161], both of which output detection windows for salient objects. We also evaluate a recent CNN-based object proposal model, MultiBox (MBox) [55, 173], which is closely related to our salient object proposal method. MBox generates 800 proposal windows, and it is optimized to localize objects of certain categories of interest (e.g., 20 object classes in PASCAL VOC [56]), regardless whether they are salient or not.

These compared methods output ranked lists of windows with confidence scores. We try different ways to compute their PR curves, such as score thresholding and rank thresholding, with or without greedy NMS, and we report their best performance. For SC and LBI, rank thresholding without NMS (i.e., output all windows above a rank) gives consistently better AP scores. Note that SC and LBI already diversify their output windows, and their confidence scores are not calibrated across images. For MBox, applying score thresholding and NMS with the IOU threshold set at 0.4 provides the best performance.

We denote our full model as SalCNN+MAP. We also evaluate two baseline methods, SalCNN+NMS and SalCNN+MMR. SalCNN+NMS generates the detections by simply applying score thresholding and greedy NMS on our proposal windows. The IOU threshold for NMS is set at 0.4, which optimizes its AP scores. SalCNN+MMR uses the maximum marginal relevance (MMR) measure to re-score the proposals [5, 21]. In our experiments, the MMR baseline follows the formulation in [21]. The MMR re-scores each proposal by iteratively selecting the proposal with maximum marginal relevance w.r.t. the previously selected proposals. The maximum marginal relevance is formulated by

$$\text{MMR} = \underset{h_i \in H \setminus H_p}{\arg\max} \left[s(h_i) - \theta \cdot \underset{h_j \in H_p}{\max} \text{IOU}(h_i, h_j) \right], \tag{6.13}$$

where H_p is the previously selected proposals. We optimize the parameter θ for the MMR baseline w.r.t. the AP score. For SalCNN, we use $\theta = 1.3$, and for MBox, we use $\theta = 0.05$. Moreover, we apply our optimization formulation (without tuning the parameters) and other baseline methods (with parameters optimized) on the raw outputs of MBox. In doing so, we can test how our MAP formulation generalizes to a different proposal model.

Evaluation Datasets We evaluate our method mainly on three benchmark salient object datasets: MSO [200], DUT-O [194], and MSRA [112].

MSO contains many background images with no salient object and multiple salient objects. Each object is annotated separately. Images in this dataset are from the testing split of the SOS dataset [200].

DUT-O provides raw bounding box annotations of salient objects from five subjects. Images in this dataset can contain multiple objects, and a single annotated bounding box sometimes covers several nearby objects. We consolidate the annotations from five subjects to generate ground truth for evaluation. To obtain a set of ground truth windows for each image, we use a greedy algorithm to merge bounding box annotations labeled by different subjects. Let $\mathbf{B} = \{B_i\}_i^n$ denote the bounding box annotations for an image. For each bounding box B_i, we calculate an overlap score:

$$S_i = \sum_{j:j \neq i} \text{IOU}(B_i, B_j).$$

Based on the overlap score, we do a greedy non-maximum-suppression with the IOU threshold of 0.5 to get a set of candidate windows. To suppress outlier annotations, a candidate window B_i is removed if there are fewer than two other windows in **B** that significantly overlap with B_i (IOU $>$ 0.5). The remaining candidates are output as the ground truth windows for the given image.

MSRA comprises 5000 images, each containing one dominant salient object. This dataset provides raw bounding boxes from nine subjects, and we consolidate these annotations in the same was as in DUT-O.

For completeness, we also report evaluation results on PASCAL VOC07 [56], which is originally for benchmarking object recognition methods. This dataset is not very suitable for our task, as it only annotates 20 categories of objects, many of which are not salient. However, it has been used for evaluating salient object detection in [60, 161]. As in [60, 161], we use all the annotated bounding boxes in VOC07 as class-agnostic annotations of salient objects.

6.3.1   Results

The PR curves of our method, baselines, and other compared methods are shown in Fig. 6.2. The full AP scores are reported in Table 6.1. Our full model SalCNN+MAP significantly outperforms previous methods on MSO, DUT-O, and MSRA. In particular, our method achieves about 15%, 34%, and 20% relative improvement in AP over the best previous method MBox+NMS on MSO, DUT-O, and MSRA, respectively. This indicates that our model generalizes well to different datasets, even though it is only trained on the SOS training set. On VOC07, our method is slightly worse than MBox+NMS. Note that VOC07 is designed for object recognition, and MBox is optimized for this dataset [55]. We find that our method usually successfully detects the salient objects in this dataset, but often misses annotated objects in the background. Sample results are shown in Fig. 6.3.

Our MAP formulation consistently improves over the baseline methods NMS and MMR across all the datasets for both SalCNN and MBox. On average, our MAP attains more than 11% relative performance gain in AP over MMR for both SalCNN and MBox, and about 12% (*resp.* 5%) relative performance gain over NMS for SalCNN (*resp.* MBox). Compared with NMS, the performance gain of our optimization method is more significant for SalCNN, because our adaptive region sampling method introduces extra proposal noise in the proposal set (see discussion in Sect. 6.2.4). The greedy NMS is quite sensitive to such noise, while our subset optimization formulation can more effectively handle it.

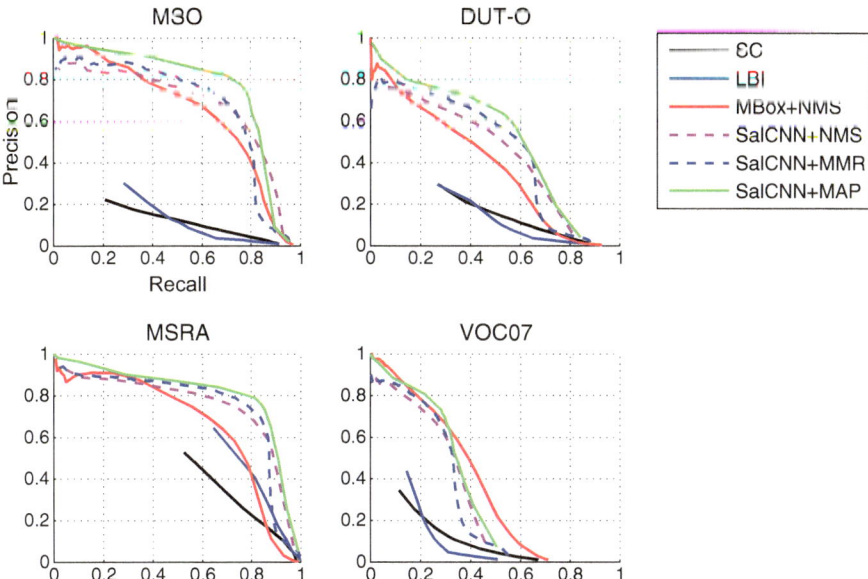

Fig. 6.2 Precision-Recall curves. Our full method SalCNN+MAP significantly outperforms the other methods on MSO, DUT-O, and MSRA. On VOC07, our method is slightly worse than MBox [173], but VOC07 is not a salient object dataset

Table 6.1 AP scores

	MSO	DUT-O	MSRA	VOC07	Avg.
SC[60]	0.121	0.156	0.388	0.106	0.194
LBI[161]	0.144	0.143	0.513	0.106	0.226
MBox[173]+NMS	0.628	0.382	0.647	0.374	0.508
MBox[173]+MMR	0.595	0.358	0.578	0.332	0.466
MBox[173]+MAP	0.644	0.412	0.676	**0.394**	0.532
SalCNN+NMS	0.654	0.432	<u>0.722</u>	0.300	0.527
SalCNN+MMR	<u>0.656</u>	<u>0.447</u>	0.716	0.301	<u>0.530</u>
SalCNN+MAP	**0.734**	**0.510**	**0.778**	<u>0.337</u>	**0.590**

The best score on each dataset is shown in bold font, and the second best is underlined

Detecting Background Images Reporting the nonexistence of salient objects is an important task by itself [186, 200]. Thus, we further evaluate how our method and the competing methods handle background/cluttered images that do not contain any salient object. A background image is implicitly detected if there is no detection output by an algorithm. Table 6.2 reports the AP score of each method in detecting background images. The AP score of our full model SalCNN+MAP is computed by varying the parameter λ specified before. For SC, LBI, MBox, and our proposal model SalCNN, we vary the score threshold to compute their AP scores.

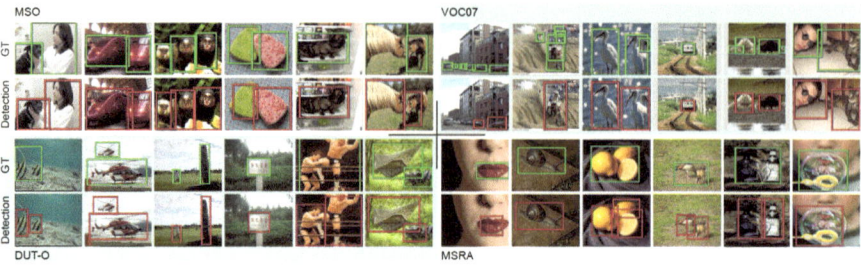

Fig. 6.3 Sample detection results of our method when $\lambda = 0.1$. In the VOC07 dataset, many background objects are annotated, but our method only detects dominant objects in the scene. In the DUT-O and MSRA datasets, some ground truth windows cover multiple objects, while our method tends to localize each object separately. Note that we are showing all the detection windows produced by our method

Table 6.2 AP scores in identifying background images on MSO

SalCNN+MAP	SalCNN	MBox+MAP	MBox	LBI	SC
0.89	0.88	0.74	0.73	0.27	0.27

Fig. 6.4 Object proposal generation performance (hit rate *vs.* average #Prop per image) on VOC07. Our MAP-based formulation further improves the state-of-the-art MBox method when #Prop is small

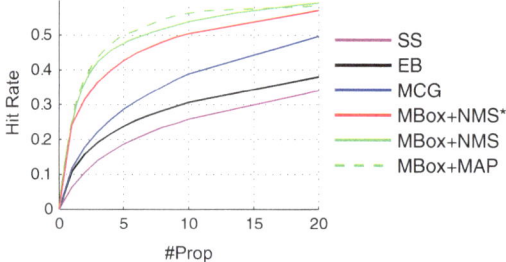

As shown in Table 6.2, the proposal score generated by SC and LBI is a poor indicator of the existence of salient objects, since their scores are not calibrated across images. MBox significantly outperforms SC and LBI, while our proposal model SalCNN achieves even better performance, which is expected as we explicitly trained our CNN model to suppress detections on background images. Our MAP formulation further improves the AP scores of SalCNN and MBox by one point.

Generating Compact Object Proposals Object proposal generation aims to attain a high hit rate within a small proposal budget [73]. When a compact object proposal set is favored for an input image (e.g., in applications like weakly supervised localization [50, 162]), how proposals are filtered can greatly affect the hit rate. In Fig. 6.4, we show that using our subset optimization formulation can help improve the hit rate of MBox [173] when the average proposal number is less than 15 (see MBox+MAP vs MBox+NMS in Fig. 6.4). The performance of MBox using rank thresholding[2] (MBox+NMS*), together with SS [180], EB [209] and MCG [5], is also displayed for comparison.

6.3.2 Component Analysis

Now we conduct further analysis of our method on the MSO dataset, to evaluate the benefits of the main components of our system.

Adaptive Region Sampling We compare our full model with two variants: the model without region sampling (w/o RS) and the model using uniform region sampling (Unif. RS) [55]. For uniform sampling, we extract five sub-windows of 70% width and height of the image, by shifting the sub-window to the four image corners and the image center. The AP scores of our full model and these two variants are displayed in Table 6.3. Besides the AP scores computed over the whole MSO dataset, we also include the results on five subsets of images for more detailed analysis: (1) 886 images with salient objects, (2) 611 images with a single salient object, (3) 275 images with multiple salient objects, (4) 404 images with all small objects, and (5) 482 images with a large object. An object is regarded as small if its bounding box occupies less than 25% area of the image. Otherwise, the object is regarded as large.

The best scores of the two variants are shown in red. The model with uniform region sampling generally outperforms the one without region sampling, especially on images with all small objects or multiple objects. However, on images with a large object, uniform region sampling worsens the performance, as it may introduce

Table 6.3 AP scores of variants of our method

		Reg. Samp.		Win. Filtering		
	Full	w/o	Unif	Rank	Score	
	Model	RS	RS	Thresh	Thresh	MMR
Overall	**0.734**	0.504	0.594	0.448	0.654	0.656
With Obj.	**0.747**	0.513	0.602	0.619	0.668	0.675
Single Obj.	**0.818**	0.676	0.671	0.717	0.729	0.721
Multi0. Obj.	**0.698**	0.338	0.540	0.601	0.609	0.620
Large Obj.	**0.859**	0.790	0.726	0.776	0.833	0.804
Small Obj.	**0.658**	0.253	0.498	0.488	0.558	0.567

Reg. Samp. refers to variants with different region sampling strategies. *Win. Filtering* refers to variants using different window filtering methods. See text for details

---

[2]Rank thresholding means outputting a fixed number of proposals for each image, which is a default setting for object proposal methods like SS, EB, and MCG, as their proposal scores are less calibrated across images.

window proposals that are only locally salient, and it tends to cut the salient object. The proposed adaptive region sampling substantially enhances the performance on all the subsets of images, yielding over 20% relative improvement on the whole dataset.

MAP-Based Subset Optimization To further analyze our subset optimization formulation, we compare our full model with three variants that use different window filtering strategies. We evaluate the rank thresholding baseline (Rank Thresh in Table 6.3) and the score thresholding baseline (Score Thresh in Table 6.3) with the greedy NMS applied. We also evaluate the maximum marginal relevance baseline (MMR in Table 6.3) as in the previous experiment.

The results of this experiment are shown in Table 6.3. Our full model consistently gives better AP scores than all of the baselines, across all subsets of images. Even on constrained images with a single salient object, our subset optimization formulation still provides 12% relative improvement over the best baseline (shown in red in Table 6.3). This shows the robustness of our formulation in handling images with varying numbers of salient objects.

6.4   Conclusion

We presented a salient object detection system for unconstrained images, where each image may contain any number of salient objects or no salient object. A CNN model was trained to produce scored window proposals, and an adaptive region sampling method was proposed to enhance its performance. Given a set of scored proposals, we presented a MAP-based subset optimization formulation to jointly optimize the number and locations of detection windows. The proposed optimization formulation provided significant improvement over the baseline methods on various benchmark datasets. Our full method outperformed the state of the art by a substantial margin on three challenging salient object datasets. Further experimental analysis validated the effectiveness of our system.

Previous methods for salient object detection either assume that there exists only a single salient object or generate a fixed number of ranked windows. The problem of unconstrained salient object detection was left unaddressed before, probably due to the lack of highly expressive image features.

Thanks to the advances in deep learning, the CNN model of our unconstrained salient object detection method is able to generate high quality object proposals with calibrated confidence scores. Combined with our proposed subset optimization formulation for bounding box filtering, our full system generates high-quality detection windows and gives significantly better performance than previous methods on unconstrained images.

Deep CNN models are able to learn highly expressive feature representations for challenging tasks, but this also comes at a cost. Training our CNN proposal generation model requires a lot of fully annotated image data. To improve the generalizability of the trained model, the training images are taken from the salient object subitizing dataset, which was carefully collected so that it covers a wide range of object categories and has a relatively balanced distribution over the number of salient objects contained in an image. Nevertheless, the data collection and annotation process for training a successful CNN model is rather expensive and laborious.

Chapter 7
Conclusion and Future Work

In this book, we have dived into classical and new tasks in visual saliency analysis. We presented methods belonging to two typical methodologies in computer vision: one based on image processing and algorithm design, and the other based on machine learning.

For the classical pixel-level saliency tasks, we proposed novel methods using image distance transform techniques. A novel distance metric, the Boolean map distance, was introduced and theoretical analysis was conducted to reveal its relation to the Minimum Barrier Distance. We proposed practical and efficient algorithms for the distance transform and demonstrate their usefulness in real-world saliency computation tasks.

Beyond the pixel level, we introduced new object-level saliency tasks, which demand more semantic understanding than the classical tasks. Advanced deep learning based methods were thus employed for these tasks. The deep learning methods are data driven, so we collected new datasets and used data augmentation techniques to improve the model training. Detailed analysis of the datasets and the learned models were presented, including model interpretation and visualization.

The whole computer vision community are embracing the deep learning era. We hope that it will allow us to study more advanced problems in visual saliency. Therefore, we conclude this book by pointing out a few directions for future work.

We first discuss a few research directions that would extend our current works.

Unconstrained Instance-Aware Salient Object Segmentation. In our work on unconstrained salient object detection, we only address the bounding box localization task for salient object detection. A more useful and more challenging goal for salient object detection is to do instance-level segmentation for salient objects. Compared with bounding boxes, segmentations can provide pixel-level localization and can be more useful in many applications like image editing. One possible approach to unconstrained salient object segmentation is to extract segmentation masks based on the bound box localization results using Grabcut-based methods [105, 144]. However, it can be challenging to properly

J. Zhang et al., *Visual Saliency: From Pixel-Level to Object-Level Analysis*, https://doi.org/10.1007/978-3-030-04831-0_7

separate overlapping objects. Another promising direction is to leverage deep CNN models to do object segmentation proposal generation, and then apply some proposal filtering method for the final output, but this approach requires additional data annotation and a careful CNN architecture design.

Category-Aware Salient Object Subitizing.   Our current subitizing model is designed to be category independent. In many applications, e.g. visual question answering [4], it will be useful to have the model to do the category-aware subitizing. A trivial approach would be training a separate subitizing model for each considered object category. A main issue here is the imbalanced distribution of the number of objects in the training images. For many categories, images in the existing datasets may lack diversity in the number of objects of the designated category. Leveraging synthetic images can be a remedy. Sharing the subitizing knowledge across different categories may also be a useful strategy for scalable model training.

We also would like to talk about some interesting visual saliency computation topics that are beyond our current research scope.

Modeling Inter-Image Context for Saliency Detection.   A majority of saliency detection methods (including the methods described in this thesis) treat each test image independently. In many scenarios, however, images are captured in groups or arranged in collections. Typical examples include videos and photo albums. Leveraging the inter-correlation of images in the same group would provide additional information for the target salient regions/objects. A related topic is co-saliency detection (see [198] for a review), which aims to detect common salient objects in a group of images. While the formulation of co-saliency detection focuses on discriminating common salient objects in an image group from uncommon salient objects, we believe that inter-image context can be useful beyond the inference of common objects. For example, a series of images about an object can provide complementary information about the object's appearance, and thus can help resolve the ambiguity in saliency detection for each individual image.

Leveraging Multimodal Signals for Saliency Detection.   Multimodal cues such as text, speech, and sound are very important factors that guide our visual attention. In real world, visual stimuli (e.g., photos and videos) are often paired with textual description (e.g., tags and captions) and audio signals. However, there are few saliency detection methods that look beyond visual information, probably due to the lack of multimodal saliency data. The problem of modeling cross-modal effects in saliency detection is also related to top-down saliency detection and priming [179]. We believe that leveraging multimodal signals can better help model and predict humans' visual attention in the real world.

Appendix A
Proof of Theorem 3.6

A.1 Preliminaries

We first introduce several concepts and results that are necessary for our proof of Theorem 3.6. In particular, we will prove a generalized version of Alexander's lemma.

A.1.1 Alexander's Lemma

The following is a version of Alexander's lemma [130].

Lemma A.1 *Let s and t be two points in the topological space $D = [0, 1]^2$, and P and Q be two disjoint closed sets in D. If there exist a path from s to t in $D \setminus P$ and a path from s to t in $D \setminus Q$, then there exists a path from s to t in $D \setminus (P \cup Q)$.*

Proof See Alexander's lemma for simply connected spaces in [88] and Theorem 8.1, p. 100 in [130].

An intuitive interpretation is shown in Fig. A.1a. As P and Q are disjoint *closed* sets, they cannot "touch" each other's boundaries. Then if neither P nor Q can block s from t, $P \cup Q$ cannot, either.

The following result is a generalization of Lemma A.1. An interpretation is shown in Fig. A.1b.

Theorem A.2 *Consider the topological space $D = [0, 1]^2$. Let $S \subset D$ be a connected set, $t \in D$ a point, and P and Q two disjoint closed sets in D. If there exist a path from S to t in $D \setminus P$ and a path from S to t in $D \setminus Q$, then there exists a path from S to t in $D \setminus (P \cup Q)$.*

© Springer Nature Switzerland AG 2019
J. Zhang et al., *Visual Saliency: From Pixel-Level to Object-Level Analysis*,
https://doi.org/10.1007/978-3-030-04831-0

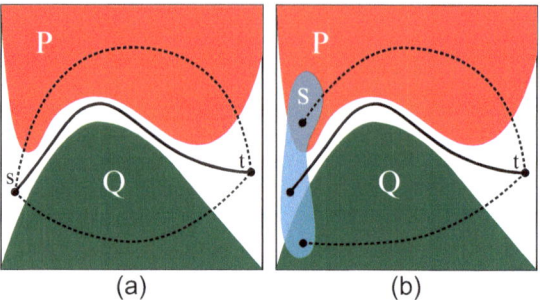

Fig. A.1 An illustration of Lemma A.1 and Theorem A.2. P and Q are two disjoint closed sets in $[0, 1]^2$. In (**a**), there is a path from s to t which does not meet P, and a path from s to t which does not meet Q (see the two dash lines). Then there exists a path (the solid line) from s to t which does not meet $P \cup Q$. In (**b**), a generalized scenario is shown, where the seed set S is not necessarily singleton, but a connected set. If S and t can be connected by paths that do not meet P and Q, respectively, then S and t can be connected by a path that does not meet $P \cup Q$

Proof We will first modify P and Q while keeping the preceding assumptions about P and Q unchanged.

Step 1 Let $\overline{P}$ denote $D \setminus P$ and $C^t_{\overline{P}} \subset \overline{P}$ denote the connected component of $\overline{P}$ that contains t. Then we change P and Q to P_1 and Q_1, respectively, s.t.

$$P_1 = P \cup \left(\overline{P} \setminus C^t_{\overline{P}} \right),$$

$$Q_1 = Q \setminus \left(\overline{P} \setminus C^t_{\overline{P}} \right).$$

Step 2 Following the same notations, we make similar changes to P_1 and Q_1:

$$Q_2 = Q_1 \cup \left(\overline{Q}_1 \setminus C^t_{\overline{Q}_1} \right),$$

$$P_2 = P_1 \setminus \left(\overline{Q}_1 \setminus C^t_{\overline{Q}_1} \right).$$

Figure A.2 gives an example to show how the above steps work. In this example, each step fills the holes inside one of the sets. Generally, as we shall prove in the following, these two steps make the complement of $P_2 \cup Q_2$ a path connected set that contains t.

We now show that the assumptions for P and Q are inherited by P_2 and Q_2. For this purpose, it suffices to show that these assumptions hold for P_1 and Q_1, because step 2 is a symmetric operation of step 1.

First, we show P_1 and Q_1 are two disjoint closed set in D. It is straightforward to see that P_1 and Q_1 are disjoint. $D = [0, 1]^2$ is locally connected, and $\overline{P} = D \setminus P$ is open in D. Each connected component of an open set in a locally connected space is open in that space [128] (Theorem 25.3 at p. 161). It follows that $\overline{P} \setminus C^t_{\overline{P}}$ is an

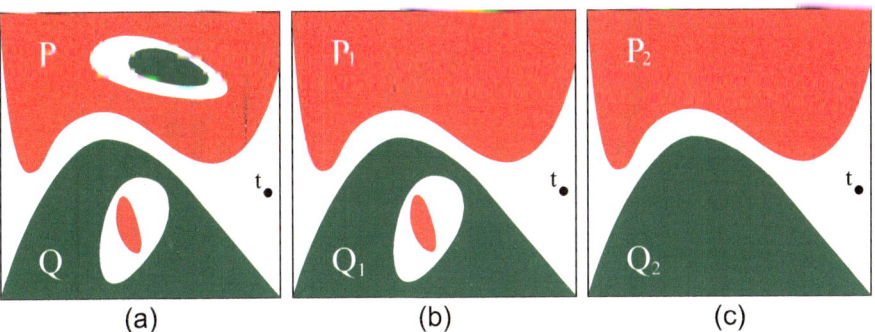

Fig. A.2 An example to show the effects of the two modification steps. In (**a**), the set shown in red, P, and the set shown in green, Q, are two disjoint closed set in $[0, 1]^2$. After step 1, the hole inside P is filled (see P_1 in (**b**)). Similarly, after step 2, the hole inside Q_1 is filled (see Q_2 in (3)). Generally, these two steps make the complement of $P_2 \cup Q_2$ a path connected set that contains t

open subset in D, because $\overline{P} \setminus C_{\overline{P}}^t$ is the union of all the components of $\overline{P}$ except $C_{\overline{P}}^t$. Thus, Q_1 is closed in D. Furthermore, it is easy to see that $P_1 = D \setminus C_{\overline{P}}^t$, so P_1 is closed in D.

Second, we show there exist a path from S to t in $D \setminus P_1$ and a path from S to t in $D \setminus Q_1$. As discussed above, $D \setminus P_1 = C_{\overline{P}}^t$. Recall that $C_{\overline{P}}^t$ is the connected component of $D \setminus P$ that contains t. Therefore, the path from S to t in $D \setminus P$ must be included in $D \setminus P_1$. Moreover, the path from S to t in $D \setminus Q$ is included in $D \setminus Q_1$ for $D \setminus Q \subset D \setminus Q_1$.

Similarly, P_2 and Q_2 have all the properties discussed above. Now we show that $D \setminus (P_2 \cup Q_2)$ is path connected. It is easy to verify the following facts:

$$P \cup Q \subset P_2 \cup Q_2, \tag{A.1}$$

$$\overline{P}_1 = C_{\overline{P}}^t, \tag{A.2}$$

$$P_2 = P_1 \cap C_{\overline{Q}_1}^t, \tag{A.3}$$

$$\overline{Q}_2 = C_{\overline{Q}_1}^t. \tag{A.4}$$

Then we have

$$D \setminus (P_2 \cup Q_2) = \overline{P_2 \cup Q_2}$$

$$= \overline{P}_2 \cap \overline{Q}_2$$

$$= \overline{P_1 \cap C_{\overline{Q}_1}^t} \cap C_{\overline{Q}_1}^t$$

$$= \left(\overline{P}_1 \cup \overline{C}_{\overline{Q}_1}^t \right) \cap C_{\overline{Q}_1}^t$$

$$= \overline{P}_1 \cap \overline{Q}_2$$

Therefore, for each $x \in D \setminus (P_2 \cup Q_2)$, we have $x \in \overline{P}_1 = C_{\overline{P}}^t$ (see Eq. A.2). Note that every open connected set in a locally path connected space is path connected [128] (Exercise 4 at p. 162). Therefore, $\overline{P}_1$ is also path connected, and x and t are path connected in $\overline{P}_1$. It follows that there exists a path joining x and t in $\overline{P}_2$, because $\overline{P}_1 \subset \overline{P}_2$. Similarly, $x \in D \setminus (P_2 \cup Q_2) \Rightarrow x \in \overline{Q}_2 = C_{\overline{Q}_1}^t$ (see Eq. A.4), so there exists a path joining x and t in $\overline{Q}_2$. Recall that P_2 and Q_2 are two disjoint closed sets in D. According to Lemma A.1, there exists a path joining x and t in $D \setminus (P_2 \cup Q_2)$. Thus, $D \setminus (P_2 \cup Q_2)$ is path connected.

Since S is connected, there exists a point $s \in S$ contained in $D \setminus (P_2 \cup Q_2)$. To see this, recall that there exist a path in $\overline{P}_2$ joining S and t and a path in $\overline{Q}_2$ joining S and t, so $S \setminus P_2 \neq \emptyset$ and $S \setminus Q_2 \neq \emptyset$. If $S \subset P_2 \cup Q_2$, then S can be divided into two disjoint nonempty closed sets in S ($S \cap P_2$ and $S \cap Q_2$), contradicting the assumption that S is connected.

$D \setminus (P_2 \cup Q_2)$ is path connected, so there is a path from s to t in $D \setminus (P_2 \cup Q_2)$. This path must be included in $D \setminus (P \cup Q)$, because $(P \cup Q) \subset (P_2 \cup Q_2)$. Thus, there is a path from S to t in $D \setminus (P \cup Q)$.

A.1.2   Hyxel and Supercover

The following concepts of *hyxel* and *supercover* are for k-D digital images. A k-D *digital image* $\mathcal{I}$ can be represented by a pair $(V, \mathcal{F})$, where $V \subset \mathbb{Z}^k$ is a set of regular grid points, $\mathcal{F} : V \rightarrow \mathbb{R}$ is a function that maps each grid point $x \in V$ to a real value $I(x)$. Moreover, we assume k-D digital images are rectangular, i.e. $V = \prod_{i=1}^{k} \{1, \cdots, n_i\}$. In particular, we are interested in the case when $k = 2$.

Next, we introduce *hyxel* [18, 43], a useful concept in digital topology.

Definition A.3 A *hyxel* H_x for a grid point x in a k-D image $\mathcal{I} = (V, \mathcal{F})$ is a unit k-D cube centered at t, i.e.

$$H_x = \left[x_1 - \frac{1}{2}, x_1 + \frac{1}{2}\right] \times \cdots \times \left[x_k - \frac{1}{2}, x_k + \frac{1}{2}\right],$$

where $x = (x_1, \cdots, x_k) \in V \subset \mathbb{Z}^k$. Hyxels are the generalization of pixels in the 2-D images.

Since hyxels and grid points in $\mathbb{Z}^k$ have a one-to-one correspondence, the connectivity of a set of hyxels can be induced by the adjacency relationship between grid points.

Definition A.4 A sequence of hyxels is an n_{adj}-*path* if the corresponding sequence of grid points is a path in terms of n-adjacency. A set of hyxels is n_{adj}-*connected* if any pair of hyxels in this set are connected by an n_{adj}-path contained in this set.

Definition A.5 The *supercover* $\mathcal{S}(M)$ of a point set M in $\mathbb{R}^k$ is the set of all hyxels that meet M.

The following is a result given in [18].

Lemma A.6 *Let $M \subset \mathbb{R}^k$ be a connected set. Then $\mathcal{S}(M)$ is $2k_{adj}$-connected.*

Proof See [18, 43]. Note that $2k_{adj}$-connectivity is equivalent to the concept of $(k-1)$-connectivity defined in [18].

A.2   Proof of Theorem 3.6

Proof First we need to show that in the continuous setting $\tilde{\mathcal{I}} = (D, \tilde{\mathcal{F}})$, where $D \subset [0,1]^2$, $\tilde{\mathcal{F}} : D \to \mathbb{R}$ is continuous, and the seed set $\tilde{S}$ is connected, $\varphi_{\tilde{\mathcal{F}}}(\tilde{S}, t)$ and $d_{\tilde{\mathcal{F}}}(\tilde{S}, t)$ are equivalent (compare Theorem 1 in [166]).

As discussed in [166], in the continuous setting, the definition of $\varphi_{\tilde{\mathcal{F}}}(\tilde{S}, t)$ and $d_{\tilde{\mathcal{F}}}(\tilde{S}, t)$ becomes

$$\varphi_{\tilde{\mathcal{F}}}(\tilde{S}, t) = h_{\tilde{\mathcal{F}}}^{+}(\tilde{S}, t) - h_{\tilde{\mathcal{F}}}^{-}(\tilde{S}, t)$$

$$= \inf_{\pi \in \Pi_{\tilde{S},t}} \beta_{\tilde{\mathcal{F}}}^{+}(\pi) - \sup_{\pi' \in \Pi_{\tilde{S},t}} \beta_{\tilde{\mathcal{F}}}^{-}(\pi'), \tag{A.5}$$

$$d_{\tilde{\mathcal{F}}}(\tilde{S}, t) = \inf_{\pi \in \Pi_{\tilde{S},t}} \beta_{\tilde{\mathcal{F}}}(\pi). \tag{A.6}$$

In order to prove $d_{\tilde{\mathcal{F}}}(\tilde{S}, t) = \varphi_{\tilde{\mathcal{F}}}(\tilde{S}, t)$, we need to show that for any $\epsilon > 0$, there exists a path $\pi \in \Pi_{\tilde{S},t}$ such that

$$h_{\tilde{\mathcal{F}}}^{-}(\tilde{S}, t) - \epsilon < \min_{x} \tilde{\mathcal{F}}(\pi(x)) \le \max_{x} \tilde{\mathcal{F}}(\pi(x)) < h_{\tilde{\mathcal{F}}}^{+}(\tilde{S}, t) + \epsilon. \tag{A.7}$$

Equivalently, we need to show there exists a path $\pi \in \Pi_{\tilde{S},t}$ in $D \setminus (P \cup Q)$, where $P = \{x \in D : \tilde{F}(x) \le h_{\tilde{\mathcal{F}}}^{-}(\tilde{S}, t) - \epsilon\}$ and $Q = \{x \subset D : \tilde{F}(x) \ge h_{\tilde{\mathcal{F}}}^{+}(\tilde{S}, t) + \epsilon\}$. It is easy to see that P and Q are two disjoint closed set in D, and there exist paths from $\tilde{S}$ to t in $D \setminus P$ and $D \setminus Q$, respectively (see the definition of $h_{\tilde{\mathcal{F}}}^{+}(\tilde{S}, t)$ and $h_{\tilde{\mathcal{F}}}^{-}(\tilde{S}, t)$ in Eq. A.5). According to Theorem A.2, Eq. A.7 is proved, and thus $\varphi_{\tilde{\mathcal{F}}}(\tilde{S}, t) = d_{\tilde{\mathcal{F}}}(\tilde{S}, t)$ in the continuous setting.

Returning to the proof of Theorem 3.8, we do the same trick as in [166] to translate the digital version of the problem to the continuous setting. We can get a continuous image $\tilde{\mathcal{I}} = (D, \tilde{\mathcal{F}})$ by bilinearly interpolating the digital image $\mathcal{I} = (V, \mathcal{F})$ Note that the seed set $\tilde{S}$ in $\tilde{\mathcal{I}}$ is the union of the hyxels (pixels) of the 4_{adj}-connected seeds in $\mathcal{I}$. Thus, $\tilde{S}$ is connected in D. According to Eq. A.7, for any $\epsilon > 0$, there exists a path $\pi' \in \Pi_{\tilde{S},t}$ such that

$$h_{\widetilde{\mathcal{F}}}^{-}(\widetilde{S}, t) - \epsilon < \min_x \widetilde{\mathcal{F}}(\pi'(x)) \le \max_x \widetilde{\mathcal{F}}(\pi'(x)) < h_{\widetilde{\mathcal{F}}}^{+}(\widetilde{S}, t) + \epsilon. \qquad \text{(A.8)}$$

Then the supercover voxelization of π' is applied. According to Lemma A.6, the resultant pixel set includes a 4_{adj}-path π that joins S and t. Because $|\mathcal{F}(x) - \widetilde{\mathcal{F}}(y)| \le \varepsilon_{\mathcal{I}}$ for any $y \in D$ that is covered by hyxel (pixel) H_x, we have

$$h_{\widetilde{\mathcal{F}}}^{-}(\widetilde{S}, t) - \varepsilon_{\mathcal{I}} - \epsilon < \min_x \mathcal{F}(\pi(x)) \le \max_x \mathcal{F}(\pi(x)) < h_{\widetilde{\mathcal{F}}}^{+}(\widetilde{S}, t) + \varepsilon_{\mathcal{I}} + \epsilon. \qquad \text{(A.9)}$$

Similarly, we have

$$h_{\widetilde{\mathcal{F}}}^{-}(\widetilde{S}, t) - \varepsilon_{\mathcal{I}} - \epsilon < h_{\mathcal{F}}^{-}(S, t) \le h_{\mathcal{F}}^{+}(S, t) < h_{\widetilde{\mathcal{F}}}^{+}(\widetilde{S}, t) + \varepsilon_{\mathcal{I}} + \epsilon. \qquad \text{(A.10)}$$

As $\epsilon \to 0$, Eq. A.9 indicates

$$d_{\mathcal{F}}(S, t) \le \varphi_{\widetilde{\mathcal{F}}}(\widetilde{S}, t) + 2\varepsilon_{\mathcal{I}}, \qquad \text{(A.11)}$$

and Eq. A.10 indicates

$$\varphi_{\mathcal{F}}(S, t) \le \varphi_{\widetilde{\mathcal{F}}}(\widetilde{S}, t) + 2\varepsilon_{\mathcal{I}}. \qquad \text{(A.12)}$$

Each discrete 4_{adj}-path on the digital image $\mathcal{I}$ has a continuous counterpart in $\widetilde{\mathcal{I}}$, which is constructed by linking the line segments between consecutive pixels on the discrete path. It is easy to see that the values on this continuous path are the linear interpolations of the values on the discrete path, for $\widetilde{\mathcal{I}}$ is obtained by bilinearly interpolating $\mathcal{I}$. Thus, $d_{\widetilde{\mathcal{F}}}(\widetilde{S}, t) \le d_{\mathcal{F}}(S, t)$ and $\varphi_{\widetilde{\mathcal{F}}}(\widetilde{S}, t) \le \varphi_{\mathcal{F}}(S, t)$. Moreover, recall that $\varphi_{\widetilde{\mathcal{F}}}(\widetilde{S}, t) = d_{\widetilde{\mathcal{F}}}(\widetilde{S}, t)$, and $\varphi_{\mathcal{F}}(S, t)$ is a lower bound of $d_{\mathcal{F}}(S, t)$ [166]. Then we have

$$\varphi_{\widetilde{\mathcal{F}}}(\widetilde{S}, t) \le \varphi_{\mathcal{F}}(S, t) \le d_{\mathcal{F}}(S, t) \le \varphi_{\widetilde{\mathcal{F}}}(\widetilde{S}, t) + 2\varepsilon_{\mathcal{I}}. \qquad \text{(A.13)}$$

It immediately follows that

$$0 \le d_{\mathcal{F}}(S, t) - \varphi_{\mathcal{F}}(S, t) \le 2\varepsilon_{\mathcal{I}}. \qquad \text{(A.14)}$$

Remark A.7 The above result can be easily generalized to k-D images. Its proof is analogous to the ones in [40, 166]. The basic idea is to reduce the problem in the high dimension to the 2-D case via homotopy [166].

Appendix B
Proof of Lemma 4.2

We provide the proof for the error bound result presented in Lemma 4.2. We start with a result about the distance transform on general graphs, which shows sufficient conditions for a locally equilibrial path map to be optimal (Sect. B.1). Then, we give the proof for our error bound result.

B.1 Distance Transform on Graph

We start with a result about the distance transform of general graphs. A graph $G = (V, E)$ is characterized by a vertex set V and an edge set E. If $(v_1, v_2) \in E$, then there is an edge from v_1 to v_2. A path on G is a sequence of vertices $\langle v_0, \cdots, v_n \rangle$, where $(v_{i-1}, v_i) \in E$ for $i = 1, \cdots, n$. The graph under consideration can be directed or undirected.

Let Π_G denote the path set on G. For a distance cost function $\mathcal{F} : \Pi_G \to \mathbb{R}^+$, without loss of generality, we assume that $\mathcal{F}$ obeys the following condition regarding a seed set S:

$$\mathcal{F}(\pi) = \begin{cases} 0 & \pi = \langle t \rangle, t \in S \\ +\infty & \pi \text{ does not start from } S. \end{cases} \tag{B.1}$$

Definition B.1 A path map $\mathcal{P}$ of a graph $G = (V, E)$ is a map that records a path $\mathcal{P}(t)$ for each vertex t on the graph. Given a seed set S, we say $\mathcal{P}$ is in its *equilibrium* state in terms of a distance cost function $\mathcal{F}$, if $\mathcal{P}(t) = \langle t \rangle, \forall t \in S$, and

$$\mathcal{F}(\mathcal{P}(t)) \leq \min_{r:(r,t)\in E} \mathcal{F}(\mathcal{P}(r) \cdot \langle r, t \rangle), \forall t \in V \setminus S. \tag{B.2}$$

© Springer Nature Switzerland AG 2019
J. Zhang et al., *Visual Saliency: From Pixel-Level to Object-Level Analysis*,
https://doi.org/10.1007/978-3-030-04831-0

In other words, if a path map is in an equilibrium state, no local update will further reduce the distance cost of any vertex. Note that there can be more than one equilibrial path map.

Definition B.2 For a graph **G**, given a seed set S and a distance function $\mathcal{F}$, a path map $\mathcal{P}$ is *optimal*, if

$$\mathcal{F}(\mathcal{P}(t)) = \min_{\pi \in \Pi_t} \mathcal{F}(\pi), \forall t \in V. \tag{B.3}$$

where Π_t is the set of paths that end at vertex t.

Solving a distance transform problem can be reduced to finding an optimal (shortest) path map for a graph and a given seed set.

Next, we introduce two properties for a distance cost function.

Definition B.3 (Non-decreasing Property) For a given graph and a seed set, if $\mathcal{F}(\pi \cdot \tau) \geq \mathcal{F}(\pi)$ always holds, we say the distance function $\mathcal{F}$ is non-decreasing.

Definition B.4 (Reduction Property) Let π_x^* denote the optimal path for a vertex x in terms of some distance function $\mathcal{F}$. For a given graph and a seed set, we say $\mathcal{F}$ is reducible, if for any non-trivial optimal path π_t^*, and any prefix π_p of π_t^*, i.e. $\pi_t^* = \pi_p \cdot \sigma$, we have $\mathcal{F}(\pi_t^*) = \mathcal{F}(\pi_p^* \cdot \sigma)$.

Non-trivial optimal paths are the optimal paths that have more than one vertex. The *non-decreasing* property and the *reduction* property are related to the smoothness conditions proposed in [57]. See [57] for a comparison. Many popular distance functions are reducible, e.g. geodesic distance and fuzzy distance [57]. Note that these two properties of a distance function can depend on the image and the seed set.

Lemma B.5 *For a given graph and a seed set, if $\mathcal{F}$ is non-decreasing and reducible, any equilibrial path map is optimal.*

Proof Given an equilibrial path map $\mathcal{P}$, let K denote set of all the vertices whose paths on $\mathcal{P}$ are optimal, and $M = V - K$ the set of all vertices whose paths are not optimal. It is easy to see that $K \neq \emptyset$ because $S \subset K$.

Suppose $M \neq \emptyset$. Let t denote a vertex of the smallest cost in M, and π_t^* its optimal path from S to t.[1] From vertex t, by tracing back along the optimal path π_t^*, we can always find a vertex $p \in M$ on π_t^*, whose preceding vertex r on π_t^* is in K, since $S \subset K$. Note that both p and t are in M, and t has the smallest cost in M, so we have

$$\mathcal{F}(\pi_t) \leq \mathcal{F}(\pi_p), \tag{B.4}$$

---

[1]The optimal path π_t^* must be a valid path from the seed set S to t. If there is no path from S to t, then $\min_{\pi \in \Pi_t} \mathcal{F}(\pi) = +\infty$, and according to Definition B.2, $p \in K$.

where $\pi_t := \mathcal{P}(t)$ and $\pi_p := \mathcal{P}(p)$ are the paths for p and t on the given path map $\mathcal{P}$. Since the map is in its equilibrium state, we have

$$\mathcal{F}(\pi_p) \leq \mathcal{F}(\pi_r \cdot \langle r, p \rangle), \tag{B.5}$$

where $\pi_r := \mathcal{P}(r)$.

Let $\pi_{r,t}^*$ denote the part from r to t on π_t^*. The non-decreasing property of $\mathcal{F}$ indicates that

$$\mathcal{F}(\pi_r \cdot \langle r, p \rangle) \leq \mathcal{F}(\pi_r \cdot \pi_{r,t}^*), \tag{B.6}$$

since $\pi_r \cdot \langle r, p \rangle$ is a prefix of $\pi_r \cdot \pi_{r,t}^*$. Furthermore, $r \in K$, so π_r is optimal. According to the reduction property, we also have

$$\mathcal{F}(\pi_t^*) = \mathcal{F}(\pi_r \cdot \pi_{r,t}^*). \tag{B.7}$$

Then combining Eqs. B.4, B.5, B.7, and B.6, we have

$$\mathcal{F}(\pi_t) \leq \mathcal{F}(\pi_t^*), \tag{B.8}$$

which contradicts our assumption that $t \in M$, i.e. π_t is not optimal. Therefore, $M = \emptyset$, which concludes our proof.

Remark B.6 Lemma B.5 indicates that for a non-decreasing and reducible distance function, any algorithm that returns an equilibrial path map, e.g. Dijkstra algorithm, fast raster scanning and parallel updating scheme, can solve the shortest path problem exactly. This result is more general than the analysis in [57], since it does not rely on any specific algorithm.

It is easy to check that the cost function $\beta_{\mathcal{I}}$ of MBD is always non-decreasing, but generally it is not reducible (see the counter-example in [166]). In the next section, we will show a sufficient condition when $\beta_{\mathcal{I}}$ is reducible, based on which our error bound result will be proved.

B.2   Proof of Lemma 4.2

A digital image $\mathcal{I}$ can be thought of as a special vertex-weighted graph. Let V denote the set of the grid points of its hyxels. For a 2-D image, we assume $V = \{1, \cdots, W\} \times \{1, \cdots, H\}$ is a rectangular region of the integer grids. The edge set E is induced by a specific type of neighborhood adjacency. We consider 4-adjacency here.

In what follows, we will first show that on the discretized image $\widetilde{\mathcal{I}}$ (see the definition in Chap. 4, Sect. 4.1.3), the path cost function of MBD is reducible when the seed set is connected. Therefore, the equilibrial path map for $\widetilde{\mathcal{I}}$ is optimal. Then we show the optimal path map for $\widetilde{\mathcal{I}}$ gives the MBD estimation for the original image $\mathcal{I}$ with errors bounded by $\varepsilon_{\mathcal{I}}$.

Definition B.7 For an image $\mathcal{I}$, a value $u \in \mathbb{R}$ is *separating* if $A = \bigcup_{\mathcal{I}(t)<u} H_t$ and $B = \bigcup_{\mathcal{I}(t)>u} H_t$ are disjoint.

Remark B.8 Note that A and B are both finite unions of closed sets (hyxels), so they are closed. The interior of A and B must be disjoint, but their boundaries may intersect. If u is separating, then the set $C = \bigcup_{\mathcal{I}(t)=u} H_t$ is in between A and B.

Lemma B.9 *Let $\mathcal{I}$ be a 2-D image, and S be a seed set. Assuming 4-adjacency, if S is connected and $\mathcal{I}(t)$ is separating for each $t \in V$, then the MBD distance cost function $\beta_{\mathcal{I}}$ is reducible.*

Proof Let $\pi_t^* \in \Pi_t$ be a non-trivial optimal path for point t, and $U_{\pi_t^*}^- = \min_i \mathcal{I}(\pi_t^*(i))$ and $U_{\pi_t^*}^+ = \max_i \mathcal{I}(\pi_t^*(i))$, where $\mathcal{I}(\pi)$ denote the set of values on the path π. It suffices to show that for any prefix of π_t^*, say π_r such that $\pi_t^* = \pi_r \cdot \sigma$, $\pi_r^* \cdot \sigma$ is also an optimal path for t. Equivalently, we need to show $\mathcal{I}(\pi_r^*) \subset [U_{\pi_t^*}^-, U_{\pi_t^*}^+]$. In what follows, we will show that $\mathcal{I}(\pi_r^*) \subset [U_{\pi_r}^-, U_{\pi_r}^+]$, where $U_{\pi_r}^- = \min_i \mathcal{I}(\pi_r(i))$ and $U_{\pi_r}^+ = \max_i \mathcal{I}(\pi_r(i))$. Since π_r is part of π_t^*, this will immediately conclude our proof.

Suppose $\mathcal{I}(\pi_r^*) \not\subset [U_{\pi_r}^-, U_{\pi_r}^+]$. Without loss of generality, we can assume that $U_{\pi_r^*}^+ = \max_i \mathcal{I}(\pi_r^*) > U_{\pi_r}^+$. In this case, we also have $U_{\pi_r^*}^- = \min_i \mathcal{I}(\pi_r^*) > U_{\pi_r}^-$, otherwise π_r^* is not optimal. Then we show there exists a path π_r^{**} such that $\mathcal{I}(\pi_r^{**}) \subset [U_{\pi_r^*}^-, U_{\pi_r}^+]$, contradicting the fact that π_r^* is optimal.

Now we show why π_r^{**} exists under our assumptions. We do this by translating the problem to the continuous setting. Let $D \subset \mathbb{R}^2$ denote the counterpart of V in the continuous space, which is defined as $D = [1, W] \times [1, H]$.[2] Let s_1 and s_2 denote the starting point of π_r and π_r^* respectively. There is a path $\widetilde{\pi}_r$ joining s_1 and r, which is included in the interior of $H_{\pi_r} = \left(\bigcup_{x \in \pi_r} H_x \right) \cap D$ in the topological space D.[3] Similarly, we can find such $\widetilde{\pi}_r^*$ joining s_2 and r in the interior of $H_{\pi_r^*}$.

Let

$$A := \bigcup \{H_x : \mathcal{I}(x) < U_{\pi_r^*}^-\} \cap D,$$

and

$$B := \bigcup \{H_x : \mathcal{I}(x) > U_{\pi_r}^+\} \cap D.$$

It is easy to see that $\widetilde{\pi}_r^*$ does not meet A and $\widetilde{\pi}_r$ does not meet B. Note that $\mathcal{I}(r) \in [U_{\pi_r^*}^-, U_{\pi_r}^+]$ and it is separating, so A and B are disjoint closed sets.

---

[2] D is properly included in $\bigcup_{x \in V} H_x$.

[3] Note that π_r is 4-connected, so by simply linking the center points of the consecutive pixels, the resultant paths will be included in the interior of the corresponding hyxel sets.

Furthermore, $H_S = \bigcup\{H_x : x \in S\}$ is connected, for S is 4_{adj}-connected. According to Lemma A.2, there is a path $\tilde{\pi}_r^{**}$ from H_S to point r which does not meet $A \cup B$.

According to Lemma A.6, the supercover $\mathcal{S}(\tilde{\pi}_r^{**})$ forms a 4_{adj}-path π_r^{**} in V. $\mathcal{S}(\tilde{\pi}_r^{**})$ cannot contain any pixel outside the digital image because there is a 0.5 wide margin between the boundaries of D and $\bigcup_{x \in V} H_x$. Furthermore, $\mathcal{S}(\tilde{\pi}_r^{**})$ cannot contain any pixel in A or B, for $\tilde{\pi}_r^{**} \subset D \backslash (A \cup B)$. Therefore, $I(\pi_r^{**}) \subset [U_{\pi_r^*}^-, U_{\pi_r}^+]$, and we arrive at the contradiction.

Let $\varepsilon_{\mathcal{I}}$ denote the *maximum local difference* (see Definition 4.1). Then we show that $\tilde{\mathcal{I}}$, the discretized image using the discretization step $\varepsilon_{\mathcal{I}}$, has the property that each value of $\tilde{\mathcal{I}}$ is separating.

Lemma B.10 *Given an image $\mathcal{I}$, we define $\tilde{\mathcal{I}}$, such that*

$$\tilde{\mathcal{I}}(x) = \left\lfloor \frac{\mathcal{I}(x)}{\varepsilon_{\mathcal{I}}} \right\rfloor \varepsilon_{\mathcal{I}}$$

Then for each $x \in V$, $\tilde{\mathcal{I}}(x)$ is separating w.r.t. $\tilde{\mathcal{I}}$.

Proof Suppose there exist a point $x \in V$ such that $\tilde{\mathcal{I}}(x)$ is not separating on $\tilde{\mathcal{I}}$. Then there exists a pair of pixels H_a and H_b that touch each other at their boundaries, s.t. $\tilde{\mathcal{I}}(a) < \tilde{\mathcal{I}}(x)$ and $\tilde{\mathcal{I}}(b) > \tilde{\mathcal{I}}(x)$. Since $\tilde{\mathcal{I}}(x)$ and $\tilde{\mathcal{I}}(a)$ are multiples of $\varepsilon_{\mathcal{I}}$, we have $\tilde{\mathcal{I}}(x) - \tilde{\mathcal{I}}(a) \geq \varepsilon_{\mathcal{I}}$. Similarly, we have $\tilde{\mathcal{I}}(b) - \tilde{\mathcal{I}}(x) \geq \varepsilon_{\mathcal{I}}$. Therefore,

$$\tilde{\mathcal{I}}(b) - \tilde{\mathcal{I}}(a) \geq 2\varepsilon_{\mathcal{I}}. \tag{B.9}$$

Let $\tilde{\mathcal{I}}(a) = n\varepsilon_{\mathcal{I}}$. Then

$$\mathcal{I}(a) < (n+1)\varepsilon_{\mathcal{I}}, \tag{B.10}$$

$$\mathcal{I}(b) \geq (n+2)\varepsilon_{\mathcal{I}}. \tag{B.11}$$

It follows that $\mathcal{I}(b) - \mathcal{I}(a) > \varepsilon_{\mathcal{I}}$. On the other hand, because H_a and H_b are intersecting at their boundaries, a and b must be 8-adjacent. Thus, we also have $|\mathcal{I}(b) - \mathcal{I}(a)| \leq \varepsilon_{\mathcal{I}}$, and a contradiction is reached.

Theorem B.11 *Let $\mathcal{I}$ be a 4-adjacent image, and $\varepsilon_{\mathcal{I}}$ be its maximum local difference. We define $\tilde{\mathcal{I}}$, such that*

$$\tilde{\mathcal{I}}(x) = \left\lfloor \frac{\mathcal{I}(x)}{\varepsilon_{\mathcal{I}}} \right\rfloor \varepsilon_{\mathcal{I}}.$$

Given a connected seed set S in terms of 4-adjacency, let $d_{\beta_{\mathcal{I}}}(S, t)$ denote the MBD for t w.r.t. the original image $\mathcal{I}$. If $\mathcal{P}$ is an equilibrial path map for $\tilde{\mathcal{I}}$ w.r.t. $\beta_{\tilde{\mathcal{I}}}$, then for each $t \in V$,

$$\left| \beta_{\tilde{\mathcal{I}}}(\mathcal{P}(t)) - d_{\beta_{\mathcal{I}}}(S, t) \right| < \varepsilon_{\mathcal{I}}.$$

Proof According to Lemmas B.5, B.9, and B.10, the equilibrial path map $\mathcal{P}$ is an exact solution for the MBD shortest path problem on $\widetilde{\mathcal{I}}$, i.e.

$$\beta_{\widetilde{\mathcal{I}}}(\mathcal{P}(t)) = \min_{\pi \in \Pi_{S,t}} \beta_{\widetilde{\mathcal{I}}}(\pi). \tag{B.12}$$

For a path π, $\beta_{\mathcal{I}}(\pi) = U^+ - U^-$, where $U^- = \min_i \mathcal{I}(\pi(i))$ and $U^+ = \max_i \mathcal{I}(\pi(i))$. Similarly, $\beta_{\widetilde{\mathcal{I}}}(\pi) = \widetilde{U}^+ - \widetilde{U}^-$, where $\widetilde{U}^- = \min_i \widetilde{\mathcal{I}}(\pi(i))$ and $\widetilde{U}^+ = \max_i \widetilde{\mathcal{I}}(\pi(i))$. According to the definition of $\widetilde{\mathcal{I}}$, we have

$$U^+ - \varepsilon_{\mathcal{I}} < \widetilde{U}^+ \leq U^+,$$

$$U^- - \varepsilon_{\mathcal{I}} < \widetilde{U}^- \leq U^-.$$

It follows that

$$U^+ - U^- - \varepsilon_{\mathcal{I}} < \widetilde{U}^+ - \widetilde{U}^- < U^+ - U^- + \varepsilon_{\mathcal{I}}$$

$$\Rightarrow |\beta_{\widetilde{\mathcal{I}}}(\pi) - \beta_{\mathcal{I}}(\pi)| < \varepsilon_{\mathcal{I}}, \text{ for any path } \pi \tag{B.13}$$

Based on Eqs. B.12 and B.13, it is easy to see that for each $t \in V$,

$$\left| \min_{\pi \in \Pi_{S,t}} \beta_{\widetilde{\mathcal{I}}}(\pi) - \min_{\pi \in \Pi_{S,t}} \beta_{\mathcal{I}}(\pi) \right| < \varepsilon_{\mathcal{I}}$$

$$\Rightarrow \left| \beta_{\widetilde{\mathcal{I}}}(\mathcal{P}(t)) - d_{\beta_{\mathcal{I}}}(S, t) \right| < \varepsilon_{\mathcal{I}}, \forall t \in V. \tag{B.14}$$

In Chap. 4, the converged solution of FastMBD* is a distance map corresponding to an equilibrial path map for $\widetilde{\mathcal{I}}$. Thus, Lemma 4.2 is proved. It also follows that any Dijkstra-like algorithm that returns an equilibrial path map *w.r.t.* the MBD path cost function has the same error bound result if a discretization step is first applied. We conjecture that the discretization step is not necessary for the error bound to hold, but the proof seems much more challenging and is left for future work.

As discussed in [40], we can assume that the digital image $\mathcal{I}$ is a discrete sampling of an *idealized* image in the continuous domain $\mathbb{R}^2$ [41], and this idealized image is a continuous function due to the smoothing effect of the point spread function in a given imaging system. Under this assumption, $\varepsilon_{\mathcal{I}}$ will approach 0, as the sampling density of the digital image increases. Therefore, Lemma 4.2 indicates that the stable solution of FastMBD* is guaranteed to converge to the exact MBD transform of the idealized image, when the sampling density of an imaging system goes to infinity.

Appendix C
Proof of the Submodularity of Function 6.11

According to Eqs. 6.10–6.12, the objective function of the proposed optimization formulation can be represented as:

$$h(\mathbf{O}) = \sum_{i=1}^{n} \max_{x_i \in \tilde{\mathbf{O}} \cup \{0\}} w_i(x_i) - \phi|\mathbf{O}| - \frac{\gamma}{2} \sum_{i,j \in \tilde{\mathbf{O}}: i \neq j} \mathcal{K}_{ij}, \qquad (C.1)$$

where $w_i(j) \triangleq \log P(x_i = j | I)$ and $\mathcal{K}_{ij}$ is shorthand for the bounding box similarity measure $\mathcal{K}(B_i, B_j)$. $\tilde{\mathbf{O}}$ denotes the index set corresponding to the selected windows in $\mathbf{O}$.

Proposition C.1 $h(\mathbf{O})$ *is a submodular function.*

Proof Let

$$h(\mathbf{O}) = \sum_{i=1}^{n} \mathcal{A}_i(\mathbf{O}) + \phi \mathcal{B}(\mathbf{O}) + \gamma \mathcal{C}(\mathbf{O}), \qquad (C.2)$$

where

$$\mathcal{A}_i(\mathbf{O}) = \max_{x_i \in \tilde{\mathbf{O}} \cup \{0\}} w_i(x_i),$$

$$\mathcal{B}(\mathbf{O}) = -|\mathbf{O}|,$$

$$\mathcal{C}(\mathbf{O}) = -\frac{1}{2} \sum_{i,j \in \tilde{\mathbf{O}}: i \neq j} \mathcal{K}_{ij}.$$

Because ϕ and γ are non-negative, it suffices to show $\mathcal{A}_i(\mathbf{O})$, $\mathcal{B}(\mathbf{O})$, and $\mathcal{C}(\mathbf{O})$ are all submodular, since the class of submodular functions is closed under non-negative linear combinations.

© Springer Nature Switzerland AG 2019
J. Zhang et al., *Visual Saliency: From Pixel-Level to Object-Level Analysis*,
https://doi.org/10.1007/978-3-030-04831-0

Recall that $\mathbf{O} \subseteq \mathbf{B} = \{B_i\}_1^n$, where $\mathbf{B}$ is the overall window proposal set. Let X and Y denote two subsets of $\mathbf{B}$, and $X \subseteq Y$. Also, let x denote an arbitrary window proposal such that $x \in \mathbf{B} \setminus Y$.

To show a function f is submodular, we just need to prove that $f(X \cup \{x\}) - f(X) \geq f(Y \cup \{x\}) - f(Y)$ [155, p. 766].

First, $\mathcal{B}(\mathbf{O})$ is submodular because

$$\mathcal{B}(X \cup \{x\}) - \mathcal{B}(X) = -|X \cup \{x\}| + |X|$$
$$= -|Y \cup \{x\}| + |Y|$$
$$= \mathcal{B}(Y \cup \{x\}) - \mathcal{B}(Y).$$

Second, $\mathcal{C}(\mathbf{O})$ is submodular because

$$\mathcal{C}(X \cup \{x\}) - \mathcal{C}(X) = -\sum_{i \in \widetilde{X}} \mathcal{K}(B_i, B_{\widetilde{x}})$$
$$\geq -\sum_{i \in \widetilde{Y}} \mathcal{K}(B_i, B_{\widetilde{x}})$$
$$\geq \mathcal{C}(Y \cup \{x\}) - \mathcal{C}(Y),$$

where $\widetilde{X}$, $\widetilde{Y}$, and $\widetilde{x}$ are the corresponding indices of X, Y, and x w.r.t. $\mathbf{B}$. Note that $\mathcal{K}(B_i, B_{\widetilde{x}})$ is a similarity measure, and it is non-negative.

Lastly, we show that $\mathcal{A}_i(\mathbf{O})$ is submodular. Note that $\mathcal{A}_i$ is a monotone set function, so $\mathcal{A}_i(Y) \geq \mathcal{A}_i(X)$. Furthermore, $\mathcal{A}_i(X \cup \{x\}) = \max\{\mathcal{A}_i(X), \mathcal{A}_i^x\}$, where $\mathcal{A}_i^x \triangleq \mathcal{A}_i(\{x\})$. Thus,

$$\mathcal{A}_i(Y \cup \{x\}) - \mathcal{A}_i(X \cup \{x\})$$
$$= \max\{\mathcal{A}_i(Y), \mathcal{A}_i^x\} - \max\{\mathcal{A}_i(X), \mathcal{A}_i^x\}$$
$$\leq \mathcal{A}_i(Y) - \mathcal{A}_i(X).$$

It is easy to see the last inequality by checking the cases when $\mathcal{A}_i^x \leq \mathcal{A}_i(X)$, $\mathcal{A}_i(X) < \mathcal{A}_i^x \leq \mathcal{A}_i(Y)$ and $\mathcal{A}_i(Y) < \mathcal{A}_i^x$, respectively. Then it follows that

$$\mathcal{A}_i(X \cup \{x\}) - \mathcal{A}_i(X) \geq \mathcal{A}_i(Y \cup \{x\}) - \mathcal{A}_i(Y).$$

Therefore, $\mathcal{A}_i$ is submodular.

References

1. ACHANTA, R., HEMAMI, S., ESTRADA, F., AND SUSSTRUNK, S. Frequency-tuned salient region detection. In *IEEE Conference on Computer Vision and Pattern Recognition (CVPR)* (2009).
2. ALEXE, B., DESELAERS, T., AND FERRARI, V. Measuring the objectness of image windows. *IEEE Transactions on Pattern Analysis and Machine Intelligence (TPAMI) 34*, 11 (2012), 2189–2202.
3. ANORAGANINGRUM, D. Cell segmentation with median filter and mathematical morphology operation. In *International Conference on Image Analysis and Processing* (1999).
4. ANTOL, S., AGRAWAL, A., LU, J., MITCHELL, M., BATRA, D., LAWRENCE ZITNICK, C., AND PARIKH, D. Vqa: Visual question answering. In *IEEE International Conference on Computer Vision (ICCV)* (2015).
5. ARBELAEZ, P., PONT-TUSET, J., BARRON, J., MARQUES, F., AND MALIK, J. Multiscale combinatorial grouping. In *IEEE Conference on Computer Vision and Pattern Recognition (CVPR)* (2014).
6. ARTETA, C., LEMPITSKY, V., NOBLE, J. A., AND ZISSERMAN, A. Interactive object counting. In *European Conference on Computer Vision (ECCV)* (2014).
7. ATKINSON, J., CAMPBELL, F. W., AND FRANCIS, M. R. The magic number 4±0: A new look at visual numerosity judgements. *Perception 5*, 3 (1976), 327–34.
8. BA, J., MNIH, V., AND KAVUKCUOGLU, K. Multiple object recognition with visual attention. In *International Conference on Learning Representations (ICLR)* (2015).
9. BARINOVA, O., LEMPITSKY, V., AND KHOLI, P. On detection of multiple object instances using Hough transforms. *IEEE Transactions on Pattern Analysis and Machine Intelligence (TPAMI) 34*, 9 (2012), 1773–1784.
10. BAYLIS, G. C., DRIVER, J., ET AL. Shape-coding in IT cells generalizes over contrast and mirror reversal, but not figure-ground reversal. *Nature Neuroscience 4* (2001), 937–942.
11. BERG, T. L., AND BERG, A. C. Finding iconic images. In *IEEE Conference on Computer Vision and Pattern Recognition (CVPR) Workshops* (2009).
12. BORJI, A., CHENG, M.-M., JIANG, H., AND LI, J. Salient object detection: A benchmark. *IEEE Transactions on Image Processing (TIP) 24*, 12 (2015), 5706–5722.
13. BORJI, A., AND ITTI, L. Exploiting local and global patch rarities for saliency detection. In *IEEE Conference on Computer Vision and Pattern Recognition (CVPR)* (2012).
14. BORJI, A., AND ITTI, L. State-of-the-art in visual attention modeling. *IEEE Transactions on Pattern Analysis and Machine Intelligence (TPAMI) 35*, 1 (2013), 185–207.
15. BORJI, A., SIHITE, D. N., AND ITTI, L. Salient object detection: A benchmark. In *European Conference on Computer Vision (ECCV)* (2012).

© Springer Nature Switzerland AG 2019
J. Zhang et al., *Visual Saliency: From Pixel-Level to Object-Level Analysis*,
https://doi.org/10.1007/978-3-030-04831-0

16. BORJI, A., TAVAKOLI, H. R., SIHITE, D. N., AND ITTI, L. Analysis of scores, datasets, and models in visual saliency prediction. In *IEEE International Conference on Computer Vision (ICCV)* (2013).

17. BOYSEN, S. T., AND CAPALDI, E. J. *The development of numerical competence: Animal and human models.* Psychology Press, 2014.

18. BRIMKOV, V. E., ANDRES, E., AND BARNEVA, R. P. Object discretizations in higher dimensions. *Pattern Recognition Letters 23*, 6 (2002), 623–636.

19. BRUCE, N. D., AND TSOTSOS, J. K. Saliency, attention, and visual search: An information theoretic approach. *Journal of Vision 9*, 3 (2009), 5.

20. BUCHBINDER, N., FELDMAN, M., NAOR, J., AND SCHWARTZ, R. A tight linear time (1/2)-approximation for unconstrained submodular maximization. In *Foundations of Computer Science* (2012).

21. CARREIRA, J., AND SMINCHISESCU, C. CPMC: Automatic object segmentation using constrained parametric min-cuts. *IEEE Transactions on Pattern Analysis and Machine Intelligence (TPAMI) 34*, 7 (2012), 1312–1328.

22. CERF, M., HAREL, J., EINHÄUSER, W., AND KOCH, C. Predicting human gaze using low-level saliency combined with face detection. In *Advances in Neural Information Processing Systems (NIPS)* (2008).

23. CHAN, A. B., LIANG, Z.-S., AND VASCONCELOS, N. Privacy preserving crowd monitoring: Counting people without people models or tracking. In *IEEE Conference on Computer Vision and Pattern Recognition (CVPR)* (2008).

24. CHAN, A. B., AND VASCONCELOS, N. Bayesian Poisson regression for crowd counting. In *IEEE International Conference on Computer Vision (ICCV)* (2009).

25. CHANG, K.-Y., LIU, T.-L., CHEN, H.-T., AND LAI, S.-H. Fusing generic objectness and visual saliency for salient object detection. In *IEEE International Conference on Computer Vision (ICCV)* (2011).

26. CHATFIELD, K., LEMPITSKY, V., VEDALDI, A., AND ZISSERMAN, A. The devil is in the details: an evaluation of recent feature encoding methods. In *British Machine Vision Conference (BMVC)* (2011).

27. CHEN, C., TANG, H., LYU, Z., LIANG, H., SHANG, J., AND SEREM, M. Saliency modeling via outlier detection. *Journal of Electronic Imaging 23*, 5 (2014), 053023–053023.

28. CHEN, J., JIN, Q., YU, Y., AND HAUPTMANN, A. G. Image profiling for history events on the fly. In *ACM International Conference on Multimedia* (2015).

29. CHEN, L. Topological structure in visual perception. *Science 218* (1982), 699.

30. CHEN, T., CHENG, M.-M., TAN, P., SHAMIR, A., AND HU, S.-M. Sketch2photo: internet image montage. *ACM Transactions on Graphics (TOG) 28*, 5 (2009), 124.

31. CHEN, X., AND GUPTA, A. Webly supervised learning of convolutional networks. In *IEEE International Conference on Computer Vision (ICCV)* (2015).

32. CHENG, M., MITRA, N., HUANG, X., TORR, P., AND HU, S. Global contrast based salient region detection. *IEEE Transactions on Pattern Analysis and Machine Intelligence (TPAMI) 37*, 3 (March 2015), 569–582.

33. CHENG, M., ZHANG, G., MITRA, N., HUANG, X., AND HU, S. Global contrast based salient region detection. In *IEEE Conference on Computer Vision and Pattern Recognition (CVPR)* (2011).

34. CHENG, M.-M., MITRA, N. J., HUANG, X., TORR, P. H. S., AND HU, S.-M. Global contrast based salient region detection. *IEEE Transactions on Pattern Analysis and Machine Intelligence (TPAMI) 37*, 3 (2015), 569–582.

35. CHENG, M.-M., WARRELL, J., LIN, W.-Y., ZHENG, S., VINEET, V., AND CROOK, N. Efficient salient region detection with soft image abstraction. In *IEEE Conference on Computer Vision and Pattern Recognition (CVPR)* (2013).

36. CHENG, M.-M., ZHANG, Z., LIN, W.-Y., AND TORR, P. H. S. BING: Binarized normed gradients for objectness estimation at 300fps. In *IEEE Conference on Computer Vision and Pattern Recognition (CVPR)* (2014).

37. CHIA, A. Y.-S., ZHUO, S., GUPTA, R. K., TAI, Y.-W., CHO, S. Y., TAN, P., AND LIN, S. Semantic colorization with Internet images. *ACM Transactions on Graphics (TOG) 30*, 6 (2011), 156.

38. CHOI, J., JUNG, C., LEE, J., AND KIM, C. Determining the existence of objects in an image and its application to image thumbnailing. *Signal Processing Letters 21*, 8 (2014), 957–961.

39. CHUA, T.-S., TANG, J., HONG, R., LI, H., LUO, Z., AND ZHENG, Y. NUS-WIDE: A real-world web image database from National University of Singapore. In *ACM International Conference on Image and Video Retrieval* (2009).

40. CIESIELSKI, K. C., STRAND, R., MALMBERG, F., AND SAHA, P. K. Efficient algorithm for finding the exact minimum barrier distance. *Computer Vision and Image Understanding (CVIU) 123* (2014), 53–64.

41. CIESIELSKI, K. C., AND UDUPA, J. K. A framework for comparing different image segmentation methods and its use in studying equivalences between level set and fuzzy connectedness frameworks. *Computer Vision and Image Understanding (CVIU) 115*, 6 (2011), 721–734.

42. CLEMENTS, D. H. Subitizing: What is it? why teach it? *Teaching children mathematics 5* (1999), 400–405.

43. COHEN-OR, D., AND KAUFMAN, A. Fundamentals of surface voxelization. *Graphical Models and Image Processing 57*, 6 (1995), 453–461.

44. DALAL, N., AND TRIGGS, B. Histograms of oriented gradients for human detection. In *IEEE Conference on Computer Vision and Pattern Recognition (CVPR)* (2005).

45. DANIELSSON, P.-E. Euclidean distance mapping. *Computer Graphics and image processing 14*, 3 (1980), 227–248.

46. DAVIS, H., AND PÉRUSSE, R. Numerical competence in animals: Definitional issues, current evidence, and a new research agenda. *Behavioral and Brain Sciences 11*, 04 (1988), 561–579.

47. DECARLO, D., AND SANTELLA, A. Stylization and abstraction of photographs. *ACM Transactions on Graphics (TOG) 21*, 3 (2002), 769–776.

48. DEHAENE, S. *The number sense: How the mind creates mathematics.* Oxford University Press, 2011.

49. DESAI, C., RAMANAN, D., AND FOWLKES, C. C. Discriminative models for multi-class object layout. *International Journal of Computer Vision (IJCV) 95*, 1 (2011), 1–12.

50. DESELAERS, T., ALEXE, B., AND FERRARI, V. Weakly supervised localization and learning with generic knowledge. *International Journal of Computer Vision (IJCV) 100*, 3 (2012), 275–293.

51. DESIMONE, R., AND DUNCAN, J. Neural mechanisms of selective visual attention. *Annual review of neuroscience 18*, 1 (1995), 193–222.

52. DEVROYE, L. *Non-uniform random variate generation.* New York: Springer-Verlag, 1986.

53. DUAN, L., WU, C., MIAO, J., QING, L., AND FU, Y. Visual saliency detection by spatially weighted dissimilarity. In *IEEE Conference on Computer Vision and Pattern Recognition (CVPR)* (2011).

54. ERDEM, E., AND ERDEM, A. Visual saliency estimation by nonlinearly integrating features using region covariances. *Journal of vision 13*, 4 (2013), 11.

55. ERHAN, D., SZEGEDY, C., TOSHEV, A., AND ANGUELOV, D. Scalable object detection using deep neural networks. In *IEEE Conference on Computer Vision and Pattern Recognition (CVPR)* (2014).

56. EVERINGHAM, M., VAN GOOL, L., WILLIAMS, C. K. I., WINN, J., AND ZISSERMAN, A. The PASCAL Visual Object Classes Challenge 2007 (VOC2007) Results. http://www.pascal-network.org/challenges/VOC/voc2007/workshop/index.html.

57. FALCÃO, A. X., STOLFI, J., AND DE ALENCAR LOTUFO, R. The image foresting transform: Theory, algorithms, and applications. *IEEE Transactions on Pattern Analysis and Machine Intelligence (TPAMI) 26*, 1 (2004), 19–29.

58. FEIGE, U., MIRROKNI, V. S., AND VONDRAK, J. Maximizing non-monotone submodular functions. *SIAM Journal on Computing 40*, 4 (2011), 1133–1153.

59. FELZENSZWALB, P. F., GIRSHICK, R. B., MCALLESTER, D., AND RAMANAN, D. Object detection with discriminatively trained part-based models. *IEEE Transactions on Pattern Analysis and Machine Intelligence (TPAMI) 32*, 9 (2010), 1627–1645.

60. FENG, J., WEI, Y., TAO, L., ZHANG, C., AND SUN, J. Salient object detection by composition. In *IEEE International Conference on Computer Vision (ICCV)* (2011).

61. GARCIA-DIAZ, A., FDEZ-VIDAL, X. R., PARDO, X. M., AND DOSIL, R. Saliency from hierarchical adaptation through decorrelation and variance normalization. *Image and Vision Computing 30*, 1 (2012), 51–64.

62. GIRSHICK, R., DONAHUE, J., DARRELL, T., AND MALIK, J. Rich feature hierarchies for accurate object detection and semantic segmentation. In *IEEE Conference on Computer Vision and Pattern Recognition (CVPR)* (2014).

63. GIRSHICK, R., DONAHUE, J., DARRELL, T., AND MALIK, J. Rich feature hierarchies for accurate object detection and semantic segmentation. In *IEEE Conference on Computer Vision and Pattern Recognition (CVPR)* (2014), IEEE.

64. GOFERMAN, S., ZELNIK-MANOR, L., AND TAL, A. Context-aware saliency detection. *IEEE Transactions on Pattern Analysis and Machine Intelligence (TPAMI) 34*, 10 (2012), 1915–1926.

65. GOPALAKRISHNAN, V., HU, Y., AND RAJAN, D. Random walks on graphs to model saliency in images. In *IEEE Conference on Computer Vision and Pattern Recognition (CVPR)* (2009).

66. GROSS, H. J. The magical number four: A biological, historical and mythological enigma. *Communicative & integrative biology 5*, 1 (2012), 1–2.

67. GROSS, H. J., PAHL, M., SI, A., ZHU, H., TAUTZ, J., AND ZHANG, S. Number-based visual generalisation in the honeybee. *PLoS One 4*, 1 (2009), e4263.

68. GURARI, D., AND GRAUMAN, K. Visual question: Predicting if a crowd will agree on the answer. *arXiv preprint arXiv:1608.08188* (2016).

69. HAN, J., NGAN, K. N., LI, M., AND ZHANG, H.-J. Unsupervised extraction of visual attention objects in color images. *IEEE Trans. Circuits and Systems for Video Technology 16*, 1 (2006), 141–145.

70. HAN, X., SATOH, S., NAKAMURA, D., AND URABE, K. Unifying computational models for visual attention. In *INCF Japan Node International Workshop: Advances in Neuroinformatics* (2014).

71. HAREL, J., KOCH, C., AND PERONA, P. Graph-based visual saliency. In *Advances in Neural Information Processing Systems (NIPS)* (2007).

72. HEO, J.-P., LIN, Z., AND YOON, S.-E. Distance encoded product quantization. In *IEEE Conference on Computer Vision and Pattern Recognition (CVPR)* (2014).

73. HOSANG, J., BENENSON, R., AND SCHIELE, B. How good are detection proposals, really? In *BMVC* (2014).

74. HOU, X., HAREL, J., AND KOCH, C. Image signature: Highlighting sparse salient regions. *IEEE Transactions on Pattern Analysis and Machine Intelligence (TPAMI) 34*, 1 (2012), 194–201.

75. HOU, X., AND ZHANG, L. Saliency detection: A spectral residual approach. In *IEEE Conference on Computer Vision and Pattern Recognition (CVPR)* (2007).

76. HUANG, H., ZHANG, L., AND ZHANG, H.-C. Arcimboldo-like collage using internet images. *ACM Transactions on Graphics (TOG) 30*, 6 (2011), 155.

77. HUANG, L., AND PASHLER, H. A Boolean map theory of visual attention. *Psychological Review 114*, 3 (2007), 599.

78. ITTI, L., AND BALDI, P. Bayesian surprise attracts human attention. In *Advances in Neural Information Processing Systems (NIPS)* (2006).

79. ITTI, L., KOCH, C., AND NIEBUR, E. A model of saliency-based visual attention for rapid scene analysis. *IEEE Transactions on Pattern Analysis and Machine Intelligence (TPAMI) 20*, 11 (1998), 1254–1259.

80. JADERBERG, M., SIMONYAN, K., VEDALDI, A., AND ZISSERMAN, A. Synthetic data and artificial neural networks for natural scene text recognition. In *Advances in Neural Information Processing Systems (NIPS) Workshop* (2014).

81. JANSEN, B. R., HOFMAN, A. D., STRAATEMEIER, M., BERS, B. M., RAIJMAKERS, M. E., AND MAAS, H. L. The role of pattern recognition in children's exact enumeration of small numbers. *British Journal of Developmental Psychology 32*, 2 (2014), 178–194.

82. JEVONS, W. S. The power of numerical discrimination. *Nature 3*, 67 (1871), 281–282.

83. JIA, Y., SHELHAMER, E., DONAHUE, J., KARAYEV, S., LONG, J., GIRSHICK, R., GUADAR-RAMA, S., AND DARRELL, T. Caffe: Convolutional architecture for fast feature embedding. In *ACM International Conference on Multimedia* (2014).

84. JIANG, B., ZHANG, L., LU, H., YANG, C., AND YANG, M.-H. Saliency detection via absorbing Markov chain. In *IEEE International Conference on Computer Vision (ICCV)* (2013).

85. JIANG, H., WANG, J., YUAN, Z., WU, Y., ZHENG, N., AND LI, S. Salient object detection: A discriminative regional feature integration approach. In *IEEE Conference on Computer Vision and Pattern Recognition (CVPR)* (2013).

86. JUDD, T., DURAND, F., AND TORRALBA, A. A benchmark of computational models of saliency to predict human fixations. In *MIT Technical Report* (2012).

87. JUDD, T., EHINGER, K., DURAND, F., AND TORRALBA, A. Learning to predict where humans look. In *IEEE Conference on Computer Vision and Pattern Recognition (CVPR)* (2009).

88. KALLIPOLITI, M., AND PAPASOGLU, P. Simply connected homogeneous continua are not separated by arcs. *Topology and its Applications 154*, 17 (2007), 3039–3047.

89. KARPATHY, A., AND FEI-FEI, L. Deep visual-semantic alignments for generating image descriptions. In *IEEE Conference on Computer Vision and Pattern Recognition (CVPR)* (2015).

90. KAUFMAN, E., LORD, M., REESE, T., AND VOLKMANN, J. The discrimination of visual number. *The American Journal of Psychology* (1949), 498–525.

91. KAZEMZADEH, S., ORDONEZ, V., MATTEN, M., AND BERG, T. L. Referitgame: Referring to objects in photographs of natural scenes. In *Conference on Empirical Methods in Natural Language Processing (EMNLP)* (2014).

92. KELLNHOFER, P., DIDYK, P., MYSZKOWSKI, K., HEFEEDA, M. M., SEIDEL, H.-P., AND MATUSIK, W. Gazestereo3d: seamless disparity manipulations. *ACM Transactions on Graphics (TOG) 35*, 4 (2016), 68.

93. KIENZLE, W., WICHMANN, F., SCHÖLKOPF, B., AND FRANZ, M. A nonparametric approach to bottom-up visual saliency. In *Advances in Neural Information Processing Systems (NIPS)* (2007).

94. KIMCHI, R., AND PETERSON, M. A. Figure-ground segmentation can occur without attention. *Psychological Science 19*, 7 (2008), 660–668.

95. KOCH, C., AND ULLMAN, S. Shifts in selective visual attention: towards the underlying neural circuitry. In *Matters of intelligence*. Springer, 1987, pp. 115–141.

96. KOFFKA, K. *Principles of Gestalt psychology*. Harcourt, Brace New York, 1935.

97. KOOTSTRA, G., NEDERVEEN, A., AND DE BOER, B. Paying attention to symmetry. In *BMVC* (2008).

98. KOURTZI, Z., AND KANWISHER, N. Representation of perceived object shape by the human lateral occipital complex. *Science 293*, 5534 (2001), 1506–1509.

99. KRÄHENBÜHL, P., AND KOLTUN, V. Geodesic object proposals. In *European Conference on Computer Vision (ECCV)* (2014).

100. KRIZHEVSKY, A., SUTSKEVER, I., AND HINTON, G. E. Imagenet classification with deep convolutional neural networks. In *Advances in neural information processing systems (NIPS)* (2012).

101. KÜMMERER, M., THEIS, L., AND BETHGE, M. Deep gaze l: Boosting saliency prediction with feature maps trained on imagenet. In *International Conference on Learning Representations (ICLR) Workshop* (2015).

102. LAROCHELLE, H., AND HINTON, G. E. Learning to combine foveal glimpses with a third-order Boltzmann machine. In *Advances in Neural Information Processing Systems (NIPS)* (2010).

103. LE MOAN, S., AND FARUP, I. Exploiting change blindness for image compression. In *IEEE International Conference on Signal-Image Technology & Internet-Based Systems (SITIS)* (2015).

104. LEE, Y. J., GHOSH, J., AND GRAUMAN, K. Discovering important people and objects for egocentric video summarization. In *IEEE Conference on Computer Vision and Pattern Recognition (CVPR)* (2012).

105. LEMPITSKY, V., KOHLI, P., ROTHER, C., AND SHARP, T. Image segmentation with a bounding box prior. In *IEEE International Conference on Computer Vision (ICCV)* (2009).

106. LEMPITSKY, V., AND ZISSERMAN, A. Learning to count objects in images. In *Advances in Neural Information Processing Systems (NIPS)* (2010).

107. LI, J., LEVINE, M. D., AN, X., XU, X., AND HE, H. Visual saliency based on scale-space analysis in the frequency domain. *IEEE Transactions on Pattern Analysis and Machine Intelligence (TPAMI) 35*, 4 (2013), 996–1010.

108. LI, X., LU, H., ZHANG, L., RUAN, X., AND YANG, M.-H. Saliency detection via dense and sparse reconstruction. In *IEEE International Conference on Computer Vision (ICCV)* (2013).

109. LI, X., URICCHIO, T., BALLAN, L., BERTINI, M., SNOEK, C. G. M., AND BIMBO, A. D. Socializing the semantic gap: A comparative survey on image tag assignment, refinement, and retrieval. *ACM Computing Surveys 49*, 1 (June 2016), 14:1–14:39.

110. LI, Y., HOU, X., KOCH, C., REHG, J. M., AND YUILLE, A. L. The secrets of salient object segmentation. In *IEEE Conference on Computer Vision and Pattern Recognition (CVPR)* (2014).

111. LIN, T.-Y., MAIRE, M., BELONGIE, S., HAYS, J., PERONA, P., RAMANAN, D., DOLLÁR, P., AND ZITNICK, C. Microsoft COCO: Common objects in context. In *European Conference on Computer Vision (ECCV)* (2014).

112. LIU, T., YUAN, Z., SUN, J., WANG, J., ZHENG, N., TANG, X., AND SHUM, H.-Y. Learning to detect a salient object. *IEEE Transactions on Pattern Analysis and Machine Intelligence (TPAMI) 33*, 2 (2011), 353–367.

113. LU, S., MAHADEVAN, V., AND VASCONCELOS, N. Learning optimal seeds for diffusion-based salient object detection. In *IEEE Conference on Computer Vision and Pattern Recognition (CVPR)* (2014).

114. LU, S., TAN, C., AND LIM, J. Robust and efficient saliency modeling from image co-occurrence histograms. *IEEE Transactions on Pattern Analysis and Machine Intelligence (TPAMI) 36*, 1 (2014), 195–201.

115. LU, Y., ZHANG, W., LU, H., AND XUE, X. Salient object detection using concavity context. In *IEEE International Conference on Computer Vision (ICCV)* (2011).

116. LUO, Y., YUAN, J., XUE, P., AND TIAN, Q. Saliency density maximization for object detection and localization. In *Asian Conference on Computer Vision (ACCV)* (2011).

117. MA, C., MIAO, Z., AND LI, M. Saliency weighted spatial pyramid representation for object recognition. In *IET International Conference on Wireless, Mobile and Multi-Media)* (2015), IET.

118. MAHADEVAN, V., AND VASCONCELOS, N. Saliency-based discriminant tracking. In *IEEE Conference on Computer Vision and Pattern Recognition (CVPR)* (2009).

119. MAIRON, R., AND BEN-SHAHAR, O. A closer look at context: From coxels to the contextual emergence of object saliency. In *European Conference on Computer Vision (ECCV)* (2014).

120. MALMBERG, F., STRAND, R., ZHANG, J., AND SCLAROFF, S. The Boolean map distance: Theory and efficient computation. In *International Conference on Discrete Geometry for Computer Imagery* (2017), Springer, pp. 335–346.

121. MANDLER, G., AND SHEBO, B. J. Subitizing: an analysis of its component processes. *Journal of Experimental Psychology: General 111*, 1 (1982), 1.

122. MARAGOS, P., AND ZIFF, R. D. Threshold superposition in morphological image analysis systems. *IEEE Transactions on Pattern Analysis and Machine Intelligence (TPAMI) 12*, 5 (1990), 498–504.

123. MARCHESOTTI, L., CIFARELLI, C., AND CSURKA, G. A framework for visual saliency detection with applications to image thumbnailing. In *IEEE International Conference on Computer Vision (ICCV)* (2009).

124. MARGOLIN, R., ZELNIK-MANOR, L., AND TAL, A. How to evaluate foreground maps? In *IEEE Conference on Computer Vision and Pattern Recognition (CVPR)* (2014).

125. MATHE, S., AND SMINCHISESCU, C. Dynamic eye movement datasets and learnt saliency models for visual action recognition. In *European Conference on Computer Vision (ECCV)* (2012).

126. MAZZA, V., TURATTO, M., AND UMILTA, C. Foreground–background segmentation and attention: A change blindness study. *Psychological Research 69*, 3 (2005), 201–210.

127. MNIH, V., HEESS, N., GRAVES, A., ET AL. Recurrent models of visual attention. In *Advances in Neural Information Processing Systems (NIPS)* (2014).

128. MUNKRES, J. R. *Topology*. Prentice Hall, 2000.

129. NATH, S. K., PALANIAPPAN, K., AND BUNYAK, F. Cell segmentation using coupled level sets and graph-vertex coloring. In *Medical Image Computing and Computer-Assisted Intervention (MICCAI)* (2006).

130. NEWMAN, M. H. A. *Elements of the topology of plane sets of points*. University Press, 1939.

131. NINASSI, A., LE MEUR, O., LE CALLET, P., AND BARBA, D. Does where you gaze on an image affect your perception of quality? applying visual attention to image quality metric. In *IEEE International Conference on Image Processing (ICIP)* (2007).

132. PAHL, M., SI, A., AND ZHANG, S. Numerical cognition in bees and other insects. *Frontiers in psychology 4* (2013).

133. PALMER, S. E. *Vision Science: Photons to Phenomenology*. The MIT press, 1999.

134. PENG, X., SUN, B., ALI, K., AND SAENKO, K. Learning deep object detectors from 3d models. In *IEEE International Conference on Computer Vision (ICCV)* (2015).

135. PETERS, R. J., IYER, A., ITTI, L., AND KOCH, C. Components of bottom-up gaze allocation in natural images. *Vision Research 45*, 18 (2005), 2397–2416.

136. PIAZZA, M., AND DEHAENE, S. From number neurons to mental arithmetic: The cognitive neuroscience of number sense. *The cognitive neurosciences, 3rd edition* (2004), 865–77.

137. PINHEIRO, P. O., LIN, T.-Y., COLLOBERT, R., AND DOLLÁR, P. Learning to refine object segments. In *European Conference on Computer Vision (ECCV)* (2016).

138. PONT-TUSET, J., ARBELAEZ, P., BARRON, J. T., MARQUES, F., AND MALIK, J. Multiscale combinatorial grouping for image segmentation and object proposal generation. *IEEE transactions on pattern analysis and machine intelligence 39*, 1 (2017), 128–140.

139. RAZAVIAN, A. S., AZIZPOUR, H., SULLIVAN, J., AND CARLSSON, S. CNN features off-the-shelf: an astounding baseline for recognition. In *IEEE Conference on Computer Vision and Pattern Recognition (CVPR), DeepVision Workshop* (2014).

140. REN, S., HE, K., GIRSHICK, R., AND SUN, J. Faster R-CNN: Towards real-time object detection with region proposal networks. In *Advances in Neural Information Processing Systems (NIPS)* (2015).

141. ROSENFELD, A., AND PFALTZ, J. L. Distance functions on digital pictures. *Pattern recognition 1*, 1 (1968), 33–61.

142. ROSENHOLTZ, R. Search asymmetries? what search asymmetries? *Perception & Psychophysics 63*, 3 (2001), 476–489.

143. ROTHE, R., GUILLAUMIN, M., AND VAN GOOL, L. Non-maximum suppression for object detection by passing messages between windows. In *ACCV* (2014).

144. ROTHER, C., KOLMOGOROV, V., AND BLAKE, A. Grabcut: Interactive foreground extraction using iterated graph cuts. In *ACM transactions on graphics (TOG)* (2004).

145. RUBIN, E. Figure and ground. *Readings in Perception* (1958), 194–203.

146. RUBINSTEIN, M., SHAMIR, A., AND AVIDAN, S. Improved seam carving for video retargeting. *ACM Transactions on Graphics (TOG) 27*, 3 (2008), 16.

147. RUJIKIETGUMJORN, S., AND COLLINS, R. T. Optimized pedestrian detection for multiple and occluded people. In *IEEE Conference on Computer Vision and Pattern Recognition (CVPR)* (2013), IEEE.

148. RUSSAKOVSKY, O., DENG, J., SU, H., KRAUSE, J., SATHEESH, S., MA, S., HUANG, Z., KARPATHY, A., KHOSLA, A., BERNSTEIN, M., BERG, A. C., AND FEI-FEI, L. Imagenet large scale visual recognition challenge, 2014.

149. RUSSAKOVSKY, O., DENG, J., SU, H., KRAUSE, J., SATHEESH, S., MA, S., HUANG, Z., KARPATHY, A., KHOSLA, A., BERNSTEIN, M., ET AL. Imagenet large scale visual recognition challenge. *arXiv preprint arXiv:1409.0575* (2014).

150. RUTISHAUSER, U., WALTHER, D., KOCH, C., AND PERONA, P. Is bottom-up attention useful for object recognition? In *IEEE Conference on Computer Vision and Pattern Recognition (CVPR)* (2004).

151. SAHA, P. K., WEHRLI, F. W., AND GOMBERG, B. R. Fuzzy distance transform: theory, algorithms, and applications. *Computer Vision and Image Understanding (CVIU) 86*, 3 (2002), 171–190.

152. SCHARFENBERGER, C., WASLANDER, S. L., ZELEK, J. S., AND CLAUSI, D. A. Existence detection of objects in images for robot vision using saliency histogram features. In *International Conference on Computer and Robot Vision* (2013).

153. SCHAUERTE, B., AND STIEFELHAGEN, R. Quaternion-based spectral saliency detection for eye fixation prediction. In *European Conference on Computer Vision (ECCV)* (2012).

154. SCHENK, F., URSCHLER, M., AIGNER, C., ROESNER, I., AICHINGER, P., AND BISCHOF, H. Automatic glottis segmentation from laryngeal high-speed videos using 3d active contours. In *Medical Image Understanding and Analysis Conference* (2014), pp. 111–116.

155. SCHRIJVER, A. *Combinatorial optimization: polyhedra and efficiency*, vol. 24. Springer Science & Business Media, 2003.

156. SEO, H. J., AND MILANFAR, P. Static and space-time visual saliency detection by self-resemblance. *Journal of vision 9*, 12 (2009), 15.

157. SERMANET, P., EIGEN, D., ZHANG, X., MATHIEU, M., FERGUS, R., AND LECUN, Y. Overfeat: Integrated recognition, localization and detection using convolutional networks. In *International Conference on Learning Representations (ICLR)* (2014).

158. SHEN, X., AND WU, Y. A unified approach to salient object detection via low rank matrix recovery. In *IEEE Conference on Computer Vision and Pattern Recognition (CVPR)* (2012).

159. SHIN, D., HE, S., LEE, G. M., WHINSTON, A. B., CETINTAS, S., AND LEE, K.-C. Content complexity, similarity, and consistency in social media: A deep learning approach. https://ssrn.com/abstract=2830377, 2016.

160. SIMONYAN, K., AND ZISSERMAN, A. Very deep convolutional networks for large-scale image recognition. In *International Conference on Learning Representations (ICLR)* (2015).

161. SIVA, P., RUSSELL, C., XIANG, T., AND AGAPITO, L. Looking beyond the image: Unsupervised learning for object saliency and detection. In *IEEE Conference on Computer Vision and Pattern Recognition (CVPR)* (2013).

162. SIVA, P., AND XIANG, T. Weakly supervised object detector learning with model drift detection. In *IEEE International Conference on Computer Vision (ICCV)* (2011), IEEE, pp. 343–350.

163. SOBRAL, A., BOUWMANS, T., AND ZAHZAH, E.-H. Double-constrained RPCA based on saliency maps for foreground detection in automated maritime surveillance. In *Advanced Video and Signal Based Surveillance (AVSS), 2015 12th IEEE International Conference on* (2015), IEEE, pp. 1–6.

164. STARK, M., GOESELE, M., AND SCHIELE, B. Back to the future: Learning shape models from 3D CAD data. In *British Machine Vision Conference (BMVC)* (2010).

165. STOIANOV, I., AND ZORZI, M. Emergence of a visual number sense in hierarchical generative models. *Nature neuroscience 15*, 2 (2012), 194–196.

166. STRAND, R., CIESIELSKI, K. C., MALMBERG, F., AND SAHA, P. K. The minimum barrier distance. *Computer Vision and Image Understanding (CVIU) 117*, 4 (2013), 429–437.

167. SUBBURAMAN, V. B., DESCAMPS, A., AND CARINCOTTE, C. Counting people in the crowd using a generic head detector. In *IEEE International Conference on Advanced Video and Signal-Based Surveillance (AVSS)* (2012).

168. SUGANO, Y., AND BULLING, A. Self-calibrating head-mounted eye trackers using egocentric visual saliency. In *Annual ACM Symposium on User Interface Software & Technology* (2015).

169. SUGANO, Y., MATSUSHITA, Y., AND SATO, Y. Appearance-based gaze estimation using visual saliency. *IEEE Transactions on Pattern Analysis and Machine Intelligence (TPAMI) 35*, 2 (2013), 329–341.

170. QUIL, D., LING, H. BEDERSON, B. B., AND JACOBS, D. W. Automatic thumbnail cropping and its effectiveness. In *ACM symposium on User interface software and technology* (2003).

171. SUN, B., AND SAENKO, K. From virtual to reality: Fast adaptation of virtual object detectors to real domains. In *British Machine Vision Conference (BMVC)* (2014).

172. SZEGEDY, C., LIU, W., JIA, Y., SERMANET, P., REED, S., ANGUELOV, D., ERHAN, D., VANHOUCKE, V., AND RABINOVICH, A. Going deeper with convolutions. In *IEEE Conference on Computer Vision and Pattern Recognition (CVPR)* (2015).

173. SZEGEDY, C., REED, S., ERHAN, D., AND ANGUELOV, D. Scalable, high-quality object detection. *arXiv preprint arXiv:1412.1441* (2014).

174. TATLER, B., BADDELEY, R., GILCHRIST, I., ET AL. Visual correlates of fixation selection: Effects of scale and time. *Vision Research 45*, 5 (2005), 643–659.

175. TAVAKOLI, H. R., RAHTU, E., AND HEIKKILÄ, J. Fast and efficient saliency detection using sparse sampling and kernel density estimation. In *Image Analysis*. Springer, 2011, pp. 666–675.

176. TOIVANEN, P. J. New geodesic distance transforms for gray-scale images. *Pattern Recognition Letters 17*, 5 (1996), 437–450.

177. TORRALBA, A., MURPHY, K. P., FREEMAN, W. T., AND RUBIN, M. A. Context-based vision system for place and object recognition. In *IEEE International Conference on Computer Vision (ICCV)* (2003).

178. TRICK, L. M., AND PYLYSHYN, Z. W. Why are small and large numbers enumerated differently? A limited-capacity preattentive stage in vision. *Psychological review 101*, 1 (1994), 80.

179. TULVING, E., AND SCHACTER, D. L. Priming and human memory systems. *Science 247*, 4940 (1990), 301–306.

180. UIJLINGS, J. R., VAN DE SANDE, K. E., GEVERS, T., AND SMEULDERS, A. W. Selective search for object recognition. *International Journal of Computer Vision (IJCV) 104*, 2 (2013), 154–171.

181. VALENTI, R., SEBE, N., AND GEVERS, T. Image saliency by isocentric curvedness and color. In *IEEE International Conference on Computer Vision (ICCV)* (2009).

182. VEDALDI, A., AND FULKERSON, B. VLFeat: An open and portable library of computer vision algorithms. http://www.vlfeat.org/, 2008.

183. VIG, E., DORR, M., AND COX, D. Large-scale optimization of hierarchical features for saliency prediction in natural images. In *IEEE Conference on Computer Vision and Pattern Recognition (CVPR)* (2014).

184. VINCENT, L. Morphological grayscale reconstruction in image analysis: applications and efficient algorithms. *IEEE Transactions on Image Processing (TIP) 2*, 2 (1993), 176–201.

185. VUILLEUMIER, P. O., AND RAFAL, R. D. A systematic study of visual extinction between- and within-field deficits of attention in hemispatial neglect. *Brain 123*, 6 (2000), 1263–1279.

186. WANG, P., WANG, J., ZENG, G., FENG, J., ZHA, H., AND LI, S. Salient object detection for searched web images via global saliency. In *IEEE Conference on Computer Vision and Pattern Recognition (CVPR)* (2012).

187. WEI, Y., WEN, F., ZHU, W., AND SUN, J. Geodesic saliency using background priors. In *European Conference on Computer Vision (ECCV)* (2012).

188. WEI, Y., WEN, F., ZHU, W., AND SUN, J. Geodesic saliency using background priors. In *European Conference on Computer Vision (ECCV)* (2012).

189. WOLFE, J. M., AND HOROWITZ, T. S. What attributes guide the deployment of visual attention and how do they do it? *Nature Reviews Neuroscience 5*, 6 (2004), 495–501.

190. XIAO, J., HAYS, J., EHINGER, K. A., OLIVA, A., AND TORRALBA, A. Sun database: Large-scale scene recognition from abbey to zoo. In *IEEE Conference on Computer Vision and Pattern Recognition (CVPR)* (2010).

191. XIONG, B., AND GRAUMAN, K. Detecting snap points in egocentric video with a web photo prior. In *European Conference on Computer Vision (ECCV)* (2014).

192. XU, K., BA, J., KIROS, R., COURVILLE, A., SALAKHUTDINOV, R., ZEMEL, R., AND BENGIO, Y. Show, attend and tell: Neural image caption generation with visual attention. *arXiv preprint arXiv:1502.03044* (2015).

193. YAN, Q., XU, L., SHI, J., AND JIA, J. Hierarchical saliency detection. In *IEEE Conference on Computer Vision and Pattern Recognition (CVPR)* (2013).

194. YANG, C., ZHANG, L., LU, H., RUAN, X., AND YANG, M.-H. Saliency detection via graph-based manifold ranking. In *IEEE Conference on Computer Vision and Pattern Recognition (CVPR)* (2013).

195. YILDIRIM, G., AND SÜSSTRUNK, S. FASA: Fast, Accurate, and Size-Aware Salient Object Detection. In *ACCV* (2014).

196. YUN, K., PENG, Y., SAMARAS, D., ZELINSKY, G. J., AND BERG, T. L. Studying relationships between human gaze, description, and computer vision. In *IEEE Conference on Computer Vision and Pattern Recognition (CVPR)* (2013).

197. ZENG, W., YANG, M., AND CUI, Z. Ultra-low bit rate facial coding hybrid model based on saliency detection. *Journal of Image and Graphics 3*, 1 (2015).

198. ZHANG, D., FU, H., HAN, J., AND WU, F. A review of co-saliency detection technique: Fundamentals, applications, and challenges. *arXiv preprint arXiv:1604.07090* (2016).

199. ZHANG, J., MA, S., SAMEKI, M., SCLAROFF, S., BETKE, M., LIN, Z., SHEN, X., PRICE, B., AND MĚCH, R. Salient object subitizing. *International Journal of Computer Vision 124*, 2 (2017), 169–186.

200. ZHANG, J., MA, S., SAMEKI, M., SCLAROFF, S., BETKE, M., LIN, Z., SHEN, X., PRICE, B., AND MĚCH, R. Salient object subitizing. In *IEEE Conference on Computer Vision and Pattern Recognition (CVPR)* (2015).

201. ZHANG, J., AND SCLAROFF, S. saliency detection: a Boolean map approach. In *IEEE International Conference on Computer Vision (ICCV)* (2013).

202. ZHANG, J., AND SCLAROFF, S. Exploiting surroundedness for saliency detection: a Boolean map approach. *IEEE Transactions on Pattern Analysis and Machine Intelligence (TPAMI) 38*, 5 (2016), 889–902.

203. ZHANG, J., SCLAROFF, S., LIN, Z., SHEN, X., PRICE, B., AND MĚCH, R. Minimum barrier salient object detection at 80 fps. In *IEEE International Conference on Computer Vision(ICCV)* (2015).

204. ZHANG, J., SCLAROFF, S., LIN, Z., SHEN, X., PRICE, B., AND MÉCH, R. Unconstrained salient object detection via proposal subset optimization. In *IEEE Conference on Computer Vision and Pattern Recognition(CVPR)* (2016).

205. ZHANG, L., TONG, M., MARKS, T., SHAN, H., AND COTTRELL, G. SUN: A Bayesian framework for saliency using natural statistics. *Journal of Vision 8*, 7 (2008), 32.

206. ZHAO, R., OUYANG, W., LI, H., AND WANG, X. Saliency detection by multi-context deep learning. In *IEEE Conference on Computer Vision and Pattern Recognition (CVPR)* (2015).

207. ZHU, J.-Y., WU, J., WEI, Y., CHANG, E., AND TU, Z. Unsupervised object class discovery via saliency-guided multiple class learning. In *IEEE Conference on Computer Vision and Pattern Recognition (CVPR)* (2012).

208. ZHU, W., LIANG, S., WEI, Y., AND SUN, J. Saliency optimization from robust background detection. In *IEEE Conference on Computer Vision and Pattern Recognition (CVPR)* (2014).

209. ZITNICK, C. L., AND DOLLÁR, P. Edge boxes: Locating object proposals from edges. In *European Conference on Computer Vision (ECCV)* (2014).

210. ZOU, W. Y., AND MCCLELLAND, J. L. Progressive development of the number sense in a deep neural network. In *Annual Conference of the Cognitive Science Society (CogSci)* (2013).